GEOMORPHOLOGICAL STUDIES IN INDIA

GEOMORPHOLOGICAL STUDIES IN INDIA

Kamla Prasad, M.A.,Ph.D.,
P.G. Department of Geography,
Vinoba Bhave University, Hazaribag, Bihar

Editor
Nageshwar Prasad, M. A., D. Litt.,
P. G. Department of Geography,
University of Burdwan, Burdwan (W. B.)

RAJESH PUBLICATIONS, NEW DELHI

Ist Published in India : 1996
ISBN : 8185891-05-2

RAJESH PUBLICATIONS
Rs. 500.00

Published by
M. L. Gupta
Rajesh Publications
1, Ansari Road, Daryaganj,
New Delhi : 110 002.

Laser Typesetting : **Anita Gupta & Sanjay Gupta**
Sanjay Enterprises, New Delhi - 110 002.

Printed at : Avon Offset Printers, New Delhi-2

EDITOR'S FOREWORD

This book entitled "Geomorphological Studies in India (A Case Study of the Rohtas Plateau)" has grown out of the Ph. D. thesis submitted in 1987 to the University of Burdwan by Dr. Kamla Prasad. This is undoubtedly a detailed and meticulous study of a karst area of India, and presents a commendable example to prove that the geomorphological studies in India have received due attention from geographers as well in recent years. Now it is a non-controversial and widely accepted feeling that geomorphic study can offer a wider understanding and a greater appreciation of the world in which we live. It attempts to satisfy our curiosity about the origin of such striking physical features as the Gilgit Canyon; the Gersoppa, Sivasamudram and Hundru warefalls; the caverns developed in and around Jabalpur, Rohtas Plateau, Cherrapunji, Kurnool and Almora; mushroom block near Jodhpur, fascinating escarpments of the Rohtas Plateau and many other features. Again, we must study the earth for a better understanding of the environment of which they are a part. An appreciation of the geomorphic history of the area may greatly aid the proper evaluation of surface materials and the configuration of the bed-rock profile. It is not an exaggeration to say that an area can be better planned for socio-economic development only with the knowledge of geomorphic aspects of that particular area.

It is with all these feelings in right directions that various aspects of geomorphology of the Indian sub-continent have been studied, mostly on regional scales ranging from drainage basins to plateau masses, mountain and piedmont zones, coastal strips, arid zones, alluvial tracts and district/province areas.

We, however, must confess that many areas are left still unexplored, unventured and partially studied. Moreover, for one reason or another, new techniques and tools are not applied in many cases. Sometimes, adequate and necessary field study is also ignored. There are limitations too. But some of the geomorphological studies in India have made great efforts to overcome these limitations. The present volume by Dr. Kamla Prasad is one of them. The Rohtas Plateau is even today inaccesible. There are great difficulties in traversing the area from most of the directions. There are rėmote areas which are either dens of dacoits or reigns of wild animals. It is why the area remained unstudied or was studied casually. Dr. Prasad has endeavoured to traverse the area on foot and tried to bring many karst features into light, though many limestone topography are still unexplored. It could not be possible by an individual effort. This work may, however, be considered as a pioneer and valuable contribution to the discipline of geomorphology particularly of karst areas. Further the work can facilitate those researchers of applied geomorphology who will aim at planning the area for different purposes.

Nageshwar Prasad
Post-Graduate Department of Geography,
The University of Burdwan
Burdwan-713104 (W.B.)

PREFACE

Being a resident of the village Sonpurwa located just on the right bank of the Son river, the author has had a long association with all the fascinating topographic features of the Rohtas Plateau from his child, though not as a keen observer till he became a student of geography at higher level. It was, however, interesting for him, in the childhood at school stage, to scale the steep scarps which come into sight the moment he comes out of his house. It was also interesting for him to cross the Plateau through**Pagdhyan Ghat** and to enter into the famous Gupteshwar Cave through intricate tunnels to worship the Lord Shiva made of stalagmites. This early association with the interesting features got impetus when he studied various topographic features under different environmental conditions at the Honours stage. It was cystallized when he undertook to write a thesis on these features for his M.A. in Geography. The interest was further deepend with the inspiration given by Dr. N. Prasad. He took immense trouble to traverse with me the area on foot (as there is no alternative) for about seven days and gave indication of many features left, to date, virgin, unexplained, overlooked and unexplored. The study area was selected to fill this gap, and the present work is the outcome of this long association.

In shaping the presents volume, efforts have been made to analyse in detail all the terrestial features of the Plateau's terrain on spatial and temporal scales and in terms of the processes which produced, sustain and transform them within these scales. The forthcoming discussion will reveal that the Rohtas Plateau is a small museum of topographic features ranging from the tectonogenetic surfaces and forms of fluvial processes to residual and karst landforms. Most of the features are fascinating owing to their peculiar forms and materials that render their existence. It is astonishing that many such captivating features have been left unrecorded, particularly of karst landforms. The present study proposes, therefore, to make geomorphological analysis of features of the Plateau as well as to make a record of all the unexplored features to fill up the gap in Indian geomorphology.

Broadly speaking, geomorphology is the study of landforms, their materials which make them and the processes that shape them. The discussion concentrates, therefore, on these three main aspects — materials, processes and forms endeavouring a complete geomorphic study of the Rohtas Plateau. The processes, one each chapter is devoted. Thus the work is divided altogether into nine chapters. Each chapter includes various aspects of discussion to achieve the goal how the landscape of the Rohtas Plateau has been shaped out. Why, where and how the plan of study has been set up, is included in the First Chapter. A critical analysis of the previous literature related to the area has also been made in this chapter.

The study of the materials and their structural characteristics has been necessiated in the context of the geomorphic evolution of the Rohtas Plateau; because these are the fundamental elements which play important in the origin and development of the landforms. Chapter 2, therefore, deals with the materials of the study area with reference to age, structural characteristics of the constituent rocks, environment of deposition of sedimentary rocks, igneous activity, folds, faults, joints and soil stratigraphy.

The central theme of this discussion is to understand temporal and spatial variations in geomorphic processes and the resulting pattern of forms in the Rohtas Plateau. It is emphasised because landforms embody the effects of both contemporary and past geomorphic processes. Geomorphic processes or events in the study area have been spatially contiguous, but have varied with regard to the nature, intensity, frequency and magnitude of occurence over different lengths of time. Chapter 3, therefore, deals with both the evolutionary and external graditional processes in greater detail.

Most of the areas in humid tropical climate evolve under the impact of running water. Therefore, in Chapter, 4 all aspects of drainage have been examined, which have ultimately helped in the understanding of the evolution of the Rohtas Plateau. An analysis of the divides of the Plateau drainage brings out salient features in relation to structure, drainage organization and the stage of denudational cycle. The study of drainage basins has been necessiated owing to the fact that drainage basin is considered to be the best unit for geomorphic investigation, particularly when we want to analyse the area in respect of the landscape evolution, processes involved therein and materials influence incurred thereupon. In a complete drainage basin one can see even the stages of landscape development. The analysis of the patterns of drainage of the Rohtas Plateau and their relations to the physiographic condition and structure have been attempted. The distribution of source heads and confluence points expresses the phases of drainage evolution and clearly depicts the nature of land and influence of structure. Similarly the analysis of stream ordering of ten drainage basins give a rough idea about the shape, size, morphology and structural characteristics of that particular basin.

The main objective of this thesis has been to bring unexplored features of the Rohtas Plateau into light and make a discussion of their genesis geomorphologically. The Rohtas Plateau is not only presenting a high land bounded by scarps, but it had numerous fascinating landforms, each speaking their own characteristics and origin. The highlands of Chotanagpur have a considerable length of escarpments on the margins of a series of Plateaus, but they are not so steep and inaccessible as in the escarpments of the Rohtas Plateau. Similarly deep-cut-valleys withalmost vertical rocky walls on both the sides are confined only to the Rohtas Plateau in the whole of the Bihar Province. Other feature like ***khohs*** and notches, waterfalls, caverns and karst features therein and hills and hillocks are really incomparable to many features of the adjoining highlands of Chotanagpur. Chapter 5 has, therefore, been devoted to discuss and analyse all these geomorphic feature of great interest.

Since relief measures are indicate of the potential energy of a drainage system, by virtue of elevation above a given datum a detailed discussion in Chapter 6 has been made by using several methods of relief representations, such as relative relief, absolute relief, average relief, dissection index, ruggedness index and their correlations.

The study of slopes in the Rohtas Plateau has been necessitated both owing to its theoretical and practical significance. We know that a particular element of slope is the product of a certain process, e.g., in the areas of fluvial processes there is high percentage of convexity over the slopes or in the arid region the concavity dominates over other elements of slope and so on. Thus if we can ascertain the amount of different elements of slope in a particular region, we can deduce therefrom the evolution of landscape, even during the geological past if the slope elements have no accordance with the present working processes. Therefore, the slope study of the Rohtas Plateau in Chapter 7 has helped in understanding the landscape evolution as well as denudational chronology.

The Rohtas Plateau itself is a geomorphic region in the highlands of Bihar. But within the delineation of the Plateau there are variations in materials, forms and processes. The Plateau top has capping of hard old red Vindhyan sandstone whereas the limestone - shale beds below are peeping through the scarps. On the foot are the debris slopes which are fringed by new alluvial tracts. Both endogenetic and exogenetic processes have evolved a variety of forms ranging from a near-flat alluvial zone to the high vertical rocky escarpments. In between these are the highly and moderately high dissected plateaus, subsided valleys, notches and ***khohs***, rsesistant outcrops in the form of ridges and hills, karst depressions and many other large and small details of topographic features. The morphometric variables like drainage texture, hydrographic network, relative relief, dissection index, average slope and many other aspects of basin morphometry clearly depict the spatial variations within the Plateau. All these variations in materials, forms and processes have been depicted in Chapter by dividing the study area into geomorphic regions.

Finally, in Chapter 9, attempts have been made to sum up the study and to present summary and review of the main findings.

Illustrations have been selected for their direct and clear impact. Many new line drawings and graphs have also been structured to this end. Tables have been reduced to the fewest possible number. Photographs have been carefully selected to illustrate precisely the subjects under discussion. A classified bibliography has been placed at the end of the discussion.

The study is new for several reasons and, in its humble way, is a contribution to the field of geomorphology. It is based on a detailed field study and intimate personal knowledge of the region acquired during the author's long stay in a peripheral area of the Plateau. To what extent has the present work succeeded is a subject to be judged by the readers.

Maithon-828207 **Kamla Prasad**

The study of slopes in the Rohtas Plateau has been of essential both owing to its theoretical and practical significance. We know that a particular element of slope is the product of a certain process, e.g., in the areas of fluvial processes there is a preponderance of convexity over the slopes or in the arid region the concavity dominates over other elements of slope and so on. Thus if we can ascertain the amount of different elements of slope in a particular region we can deduce therefrom the evolution of landscape, even during the geological past if these slope elements have the accordance with the present working processes. Therefore, the slope study of the Rohtas Plateau in Chapter 7 has helped in understanding the landscape evolution as well as denudational chronology.

The Rohtas Plateau itself is a geomorphic region in the highlands of Bihar. But within the delimitation of the Plateau there are variations in its rocks, forms and processes. The plateau top is capping of hard coloured Vindhyan sandstone whereas the limestone-shale beds below are peeping through the scarps. On the foot are the debris slopes which are fringed by the alluvial fans. Both endogenetic and exogenetic processes have evolved a variety of forms ranging from a near flat alluvial zone to the high vertical rocky escarpments, and between these are the hilly and moderately high dissected plateaus, subdued valleys, trenches and gorges, residual outcrops in the form of ridges and hills, karst depressions and many other large and small details of topographic features. The morphometric variables, like drainage texture, hydrological network, relative relief, dissection index, average slope and many other aspects of micro-topography clearly depict the spatial variations within the Plateau. All these variations in materials, forms and processes have been depicted in Chapter 8 by dividing the study area into geomorphic regions.

Finally, in Chapter 9, attempts have been made to sum up the study and to present summary and review of the major findings.

Illustrations have been selected for their direct and clear import. Many line drawings and graphs have also been appended to this end. Tables have been reduced to the lowest possible number. Photographs have been carefully selected to illustrate precisely the important features. An essential bibliography has been given at the end of the discussion.

The volume is new for several reasons and to a certain degree, it is an original contribution to geomorphology. It is based on a thorough field study and intimate personal knowledge of the terrain acquired during the author's long stay in the peripheral area of the Plateau. To what extent has the present work succeeded is a subject to be left to the readers.

Patna, [illegible] [illegible] Prasad

Acknowledgement

The author expresses his profound gratitude to Dr. Nageshwar Prasad, Reader, Department of Geography, Burdwan University, Burdwan (West Bengal) for his brotherly encouragement, friendly co-operation in the field and other works, scholarly supervision, and continual commitment of time throughout the course of this work from checking, examining and compilation of information to the summing up of this in spite of his demanding academic schedule.

Indebtedness is due to him again for his generosity shown in editing of present work and getting it published in this format. He has revised it thoroughly to see it up-to-date. Some of the maps are originally drawn by him.

The author owes deep gratitude to his teachers Dr. E. Ahmad, Dr. A. Prasad, Dr. D.D.P.Sapathi and many others of Ranchi University for inspiration to complete this task.Indebtedness is due to elder brother Sri B.C. Choudhary, B.D.O., Madanpur Block, Aurangabad for collecting Gazetteers, maps and other relevant information from different Government Offices, and for his constant encouragement during the course of the work.

The author wishes to express his deep sense of gratitude to Smt. Meena Devi for her patience during the work and for the great responsibility she took in looking after the children during his absence from Maithon.

The author expresses his sincere thanks to Smt. Sushma Prasad Arya and other members of her family for their homely co-operation during his stay at Burdwan and encouragement for the completion of this work.

The author also takes this oppurtinity to express his thanks to the university Grants Commission, New Delhi without whose financial aid, it would not have possible to go to the field with various equipments, and to bring out the volume in this form.

The author would like to extend his deep sense of gratitude to a large number of friends, Officials and public men who have helped in the achievement of this difficult task : Shri T.N. Mishra, Dy. Director, Bihar Mineral Development Corporation, Ranchi ; Sri Heera Prasad, S.D.O., Durgawati Project; Dr. M.M. Pathak, Dr. P. Mahto, Dr. V.Shiva and Dr. P.K. Sinha, B.S.K. College, Maithon; Dr. Abhimanyu Mahto, Marwari College, Ranchi; Dr. K.P. Yadav,

Bhabhua College; Sri Shambhu Prasad, S.H.M.S. College, Kumardhubi, Sri Gyan Prakash Yadav; and my villagers who gave company in the field.

Finally hearty thanks are also due to Sri M.L. Gupta, Proprietor, Rajesh Publications, New Delhi for the help and encouragement I received to get my work published.

Maithon **Kamla Prasad**

CONTENTS

Editor's Foreword	V
Preface	VII
Acknowledgement	IX
Figures & Plates	XI
Tables	XII
Chapter 1 : Introduction	1
Chapter 2 : Materials & their Structural Characteristics	16
Chapter 3 : Geomorphic Processes	36
Chapter 4 : Drainage Aspects	57
Chapter 5 : Surface Features & Karst Landscape	93
Chapter 6 : Relief Measures	120
Chapter 7 : Slope Morphology	133
Chapter 8 : Geomorphic Regions	148
Chapter 9 : Conclusion	163
Bibliography	170

FIGURES

1 : Space Relations 2
3.1 : Geological Evolution 40
3.2 : Natural Vegetation 48
4.1 : Drainage Divides & Basins 59
4.2 : Hydrographic Network 73
4.3 : Hydrographic Network 75
4.4 : Drainage Density 77
4.5 : Stream Frequency 79
4.6 : Stream Hierarchy : Ausane Basin 88
4.7 : Stream Hierarchy : Sura II Basin 89
5.1 : Surface Features 94
5.2 : Gupteshwar Cave 118
6.1 : Relative Relief 121
6.2 : Absolute Relief 124
6.3 : Average Relief 126
6.4 : Dissection Index 128
6.5 : Ruggedness Index 131
7.1 : Morphometry by Profiles (West-East) 134
7.2 : Morphometry by Profiles (South-North) 135
7.3 : Average Slope 140
7.4 : Slope Category 142
8.1 : Geomorphological Map 149
8.2 : Geomorphic Regions 150

PLATES

1 Structural Characteristics 19
2 Nature of Scarps 100
3 Waterfall 108
4 Plunge Pools & Subsided Valleys 108
5 Limestone Caves 113
6 Limestone Caves 113

TABLES

2.1 : Lithostratigraphic Subdivision of the Vindhyan Supergroup. 18
2.2 : The Semri Series 20
2.3 : The Kaimur Series 24
2.4 : Lithostratigraphy and Depositional Environment of the Vindhyan Supergroup. 26
2.5 : Environment of Deposition of the Upper Vindhyans 27
2.6 : Palaeocurrent data of the time of Vindhyan Period 28
3.1 : The Lower Vindhyan 41
3.2 : The Upper Vindhyan 42
3.3 : The Temperature Data (Indrapuri Barrage) 45
3.4 : Annual Rainfall of the Rohtas District (Blockwise) 46
3.5 : Monthly Rainfall in the Adhaura Block 47
3.6 : Natural Vegetation 49
4.1 : Drainage Basins & Basin Shape 58
4.2 : Source Heads 72
4.3 : Confluence Points 72
4.4 : Drainage Density 78
4.5 : Stream Frequency 78
4.6 : Basin Morphology of the Goptha, Kohira, Lorsi, Ausane, Koel, Durgawati, Karsotha, Dhoba, Sura I & Sura II 82-86
4.7 : Sinuosity Indexes (Mueller's Model) 91
5.1 : Notches & Khohs 98
5.2 : Distribution of Isolated Hills 106
5.3 : Waterfalls 110
5.4 : Distribution of Caves 114
6.1 : Relative Relief 122
6.2 : Absolute Relief 123
6.3 : Average Relief 125
6.4 : Dissection Index 127
6.5 : Ruggedness Index 130
7.1 : Average Slope 139
7.2 : Slope Category 141
8.1 : Geomorphic Regions 150

1

INTRODUCTION

1.1. EXTENT

The Rohtas Plateau lies between latitudes 24° 30' 35" and 25° 00' 00" N., and longitudes 83° 20' 45" and 84° 05' 10" E., covering an area of about 2340 square kilometres, a little more than one and a half of the area of the Union Territory of Delhi. The Plateau shares 31.35 per cent of the district of Rohtas in the state of Bihar in India (Fig. 1). It comprises the anchals of Adhaura, Nawhatta and Rohtas in full and Chainpur, Chenari, Sheosagar, Sasaram and Bhagwanpur in part, all anchals belonging to the district of Rohtas, which was once a part of Shahabad district (Fig.1). Bordered by the Son river on the south, south-east and east, the Plateau of the smooth boundary on the south and south east and indentation owing to notches and khohs on the north. The maximum west-east extent of the Plateau is about 71.5 kilometres and the maximum north-south width about 52.0 km. It attains its maximum length along 24° 50' North latituded and maximum width is along 83°32' 30" E. longitudes. The southern and south-eastern boundary makes roughly an arc of a circle while the northern portion presents a fiord or finger-like projections. The Plateau has maximum north-south width towards the north-west, it becomes narrower and narrower in the middle and again it has a horn-like projection towards, smoothness being rare. The perimeter of the area is 341 kilometres.

1.2. LOCATION

The Rohtas Plateau, a well defined and distinctive unit providing a wide and interesting field for geomorphological study, is located in the south-western corner of the district of Rohtas in Bihar. It is a part of the Indian Peninsular upland and forms the easternmost edge of the kaimur Plateau, the lowest of the three Vindhyan Plateaus which cover a substantial part of Central India. The Vindhyan has an interesting location in the sense that on the west it glances over a long stretch of the Kaimur Range of Pre-Cambrian and Vindhyan sediments consisting of massive stratified sandstone, intercalated with shales and limestone; on the south it descends abruptly on the wide course of the Son beyond which is the narrow alluvial section of the western south Gangetic Bihar and then a vast stretch of the Plateau country of the outer Chotanagpur of gneisses and granites; and on the north and east the Plateau steps down on the alluvial plains of south Bihar. Thus the Plateau is located between the oldest geological formation on the one side and on the other the alluvium of recent and ultra-recent origin. Similarly, the region is situated between the high range of relief and physiographic units on the west and south and the low monotonous relief on the north and east.

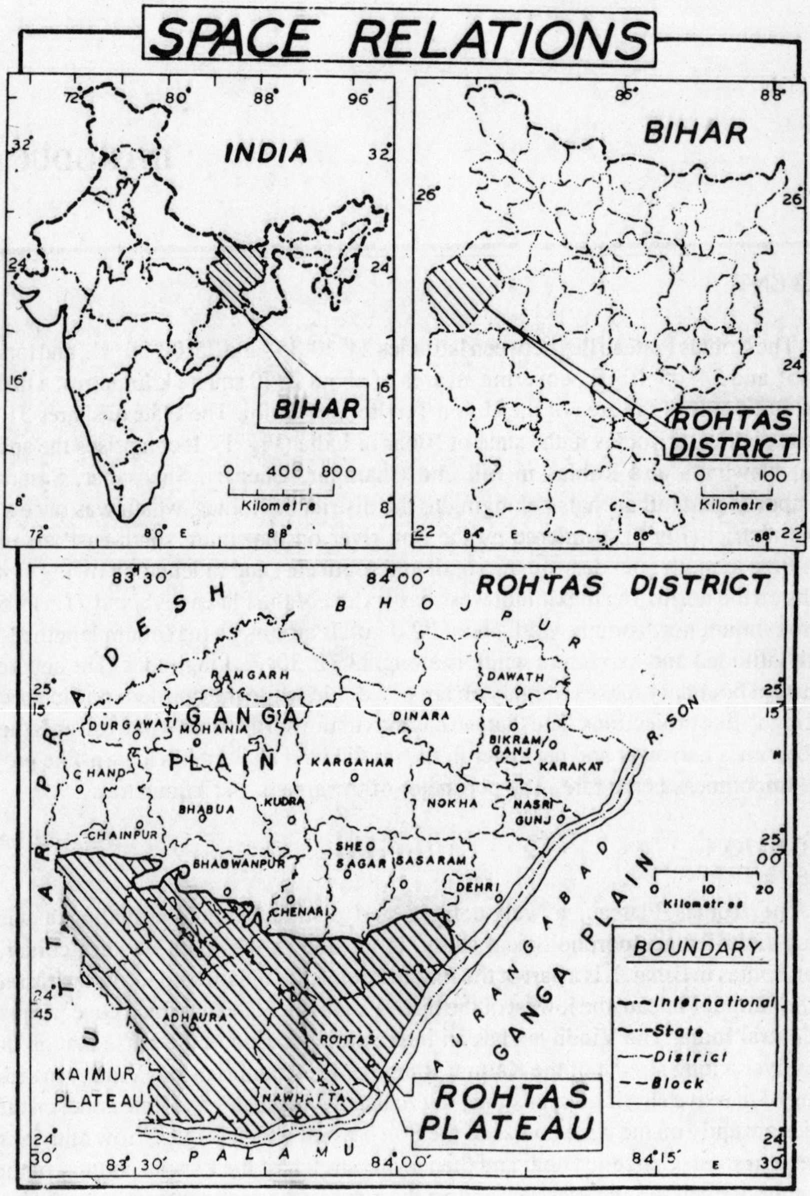
SPACE RELATIONS
INDIA
BIHAR
0 400 800
Kilometre
BIHAR
ROHTAS DISTRICT
100
Kilometre
ROHTAS DISTRICT
UTTAR PRADESH
BHOJPUR
GANGA PLAIN
RAMGARH
DURGAWATI
MOHANIA
CHAND
DINARA
DAWATH
BIKRAM GANJ
KARAKAT
KARGAHAR
R. SON
BHABUA
KUDRA
NOKHA
NASRI GUNJ
CHAINPUR
BHAGWANPUR
SHEO SAGAR
SASARAM
DEHRI
CHENARI
AURANGABAD
GANGA PLAIN
ADHAURA
ROHTAS
KAIMUR PLATEAU
NAWHATTA
PALAMU
0 10 20
Kilometres
BOUNDARY
International
State
District
Block
ROHTAS PLATEAU

FIG. 1

1.3. DELIMITATION

The study area is a **Plateau**. In geomorphology, a **drainage basin** or **catchment basin** is generally considered the best unit for investigations for several reasons. One of them is its easy demarcation by water divides on all sides. Another important aspect is related to its geomorphic evolution. A basin has **wholeness** of the stages (youth, maturity and old), through which processes work on materials to develop landforms. It may differ in **scale**, and stages may be **relative** in terms of those in other basins, but **entireness** is there. This wholeness is seldom found in a Plateau, neither is it possible. This is also for several reasons. Let us first try to define a Plateau.

Many scholars and organizations (Dana, 1885; Powell, 1896, pp. 34-64; Fennyman, 1916, p. 202; Worcester, 1948, pp. 514-15; Engeln, 1948, p. 62; Holmes, 1949, p. 414; Moore, 1963, p. 16; Encyclopedia of geomorphology, 1968, pp. 856-858; Monkhouse, 1970, p. 324; Cheng, 1971, p. 17; Frank and Raymond, 1978, p. 126; Cooke & Doornkamp, 1978, (Preface); Sanders, 1981, p. 406; Webster's New Collegiate Dictionary, 1983) have tried to define the term **plateau**.

In assence a Plateau, with its synonyms of 'table-land', 'plateform', 'elevated uplands', 'high level regions', 'highland', and 'mesa', **etc**. is defined as an extensive stretch of elevated and comparatively level land raised or dropped distinctly and sharply above or below bordering areas, at least on one side. According to the mode of origin surface forms, geographical location, climatic condition and age, a plateau comes or is used with an adjective, **e.g**. 'diastrophic' (plateau)' 'dissected', 'pediment', 'dry' or 'glaciated', and 'young' or so on. By definition, a plateau should have **some amount of elevation** and **an abrupt rise or drop** from the surface to the adjacent land, at least on one side. There is, however, no consensus on the height nor on the amount of steepness of the rise and fall. It is arbitrary and subjective; but the **tabular form** always comes to mind.

Accordingly, a plateau will seldom have, in humid regions, the wholeness of a drainage basin, as stated earlier. The edges of a plateau will be the source regions of rivulets and streams flowing in opposite directions; consequently a plateau is an area of a number of small and large drainage basins, complete as well as incomplete. Thus it becomes sometimes difficult to analyse and discuss the geomorphic evolution of a plateau through stages.

In spite of all these problems, a plateau is taken as a unit for geomorphic investigations mainly because of the fact that many plateaus of the earth are well defined and have sufficient extensive area and varied features for a geomorphological study. They can be studied in terms of materials and processes in detail. The stages of development can be determined by other methods, such as by studying small drainage basins within the plateau, measuring rate erosion and recession of tabular faces and by conducting a comparative analysis of the adjacent regions. The Rohtas Plateau, with its extensive stretch of elevated and comparatively level land raised distinctly and sharply above the bordering areas on the south, east and north, has been considered a unit for geomorphic investigations.

The delimitation of a plateau is, however, not an easy task; because abrupt rise and fall or wall-like boundary is not always available. The moment a land surface rises above the baselevel, denudation work starts functioning; consequently, the surface is subject to reduction, recession and modification in various ways. It happens even in case of those materials and forms which are resistant to erosion. Similar is the case with the Plateau of Rohtas. The forthcoming chapter (Ch. 3.1.2) makes it clear that the Plateau originated due to an uplift during the Himalayan orogeny in Tertiary times. Naturally, it might have been demarcated by a clear-cut fault-scarp, at least on the north and east. The antecedent Son river might have maintained its course on the south separating the Plateau from the rest of the highlands of Chotanagpur and causing the southern face a bold scarp. On the west, it was continued with the Kaimur Range and the Vindhyans. The bold and steep rises on the south, east and north were subject to erosion and recession after the uplift, and in course of time, the original boundary was modified. At present the southern and eastern scarps are a little far from the left bank of the Son, but the materials of scarps resemble those of the banks and the bed of the Son. It is supported also by Chhibber (1954, p. 12) when he observes that the scarps must have existed originally on the edge of the Son. Even from its present edge there has been a recession of about 4 miles, but in some cases, **e.g**. in the case of the Durgawati Nala, the recession has been to the extent of about 16 miles, while in the case of the Ausane Valley in the south it has been to the extent of at least 8 miles and now the northern and southern edges are only a little more than a mile apart'. Therefore, it was logical for the author to delimit the southern and eastern boundary to the left bank of the Son river. The hills and hillocks situated in the areas lying between the scarps and the Son, **viz.**, Katudanr, Gamharia, Tipa-Bhawnathpur, Amarkha in the south; Murli **pahari** in the south east; and Karwandia **pahari** in the north-east are but the remnants of the plateau which have been separated from the main mass by rivers and streams. Thus Son makes the southern, south-eastern and eastern boundary. Since it also forms an administrative boundary between Rohtas district and Palamu on the south and Rohtas district and Aurangabad on the east, the southern and eastern boundary coincide with the administrative boundary.

The Plateau is very much indented on the north (Fig. 5). It appears that the northerly dip of the Plateau has given rise to a number of north flowing streams which have, with their large volume of water and catchment area, notched down the north facing scarps. It also appears (Ch. 7.2) that the collapse of caverns has originated steep and bare rocky walled **khohs** which have been responsible for the high indentation of scarps. The zone which was spared by the recession of scarps has been filled up with thick alluvial deposits by the drainage system of the Ganga. It has, therefore, been difficult for the author to search out the original northerly extent of the Plateau. The author thought it logical to take the present line of scarps as northern boundary, though it is very much indented. This northern boundary of the Plateau does not, however, coincide with any administrative boundary.

As regards the western delimitation of the Plateau, it was problematic. There is no scarp and it has a continued elongation up to the Vindhyan Range in Madhya Pradesh and Rajasthan. This whole Central Range cannot be studied in detail considering a single geomorphic unit. The regionalisation was desirable. In case of the Rohtas Plateau, it was convenient to include the area belonging to **Rohtas**. At present, Rohtas is the name of a district of Bihar and its western limit makes the boundary between Bihar and Uttar Pradesh for a distance of about 78 kilometres. The belief is that it was the seat of King Rohitasva, son of Harischandra in ancient times. Rohtas fort, later occupied by Kharwars, Shershah and Akbar and now left abandoned, is making the belief sound (D.G., 1966,pp.867-870). The Rohtas Plateau is also known as **Lohitgiri** in Sanskrit literature (Quoted from Pandey, 1963, p. 49). That appears to have come either from Rohit (short form of Rohitasva), derivatively from the red look (Lohu, means blood; lohit means blood-look) of the area owing to the distribution of Kaimur red sandstone throughout the area. Some local tribals say that **Roh** in Mundari means dry and **Tas** means land; so Rohtas means dryland' (Lal, 1985, p. 7). Whatever may be the derivation, the name Rohtas is applicable to the area spread within the limits of Bihar. The author has, therefore, delimited the western boundary of the Rohtas Plateau with the State boundary between Bihar and Uttar Pradesh. Ridges and streams have justified this delimitation. From the Son in the south to the point coinciding with 24⁰ 41' N. latitudes in the north, a ridge makes the boundary which is followed by a major stream, Karamnasha up to 24⁰42' latitude. Again, a ridge is elongated up to 24⁰44' and then the rest of the boundary is made by the Guruwat and the Karamnasha rivers. Other geomorphological details such as drainage, texture, relief, slope and ruggedness differ considerably in Uttar Pradesh area from the Rohtas Plateau. Therefore, the western limit, though not a physical boundary, appears to be logical and desirable.

Thus, finally, the study area, the Rohtas Plateau is delimited by the left bank of the son on the south and east, indented scarps on the north, and the state boundary between Bihar and Uttar Pradesh on the west (Fig.1).

1.4. LANDFORMS

Within the extent of the Rohtas Plateau, delimited just in the preceding lines, there are a number of fascinating landforms both on the surface and within the strata of sedimentary rocks. The singular feature of the Plateau, which distinguishes it from the adjoining highlands of Chotanagpur, is an abrupt rise of the flattopped plateau. The highlands of Chotanagpur have a series of plateaus separated by steep scarps, but the steepness is modified here, and so they are not so fascinating as those of the Rohtas Plateau. One can see the majestic escarpment faces of the Plateau on the left bank of the Son, whose 'sheer loftiness will fail to impress but few' (Chakraborty, 1965, p. 107). On the east and north these escarpment faces have been notched through the narrow and elongated river valleys, locally called **khohs** with overhanging cliffs, waterfalls (Plate:3), dark and cool plunge pools (Plate:4) and springs, sometimes observed together at the same place. Where the base of the **Khohs** is wider, and walls are overhanging, it becomes apple-shaped. All these features together are very charming and will attract any one, particularly a student of geomorphology to deal with their evolution and characteristics.

Another fascinating feature is the presence of solution caves (Plates : 5 & 6). The caves are numerous and they are comprising tunnels of intricate patterns. They abound in stalactites, stalagmites, helictites, Karst pillars, drip curtains, lakes and other natural subterranean features which are numerous and most complex. The famous Gupteshwar Cave is one of them.

Apart from these fascinating sites, the plateau has other topographic features of geomorphic interest, such as the remarkable ruggedness and undulations of the Plateau top with saucer-like depressions at places, the hummocky surface of the boulder-pebble landform rising gradually towards the Plateau base along identifiable fluvial valleys, limestone gorges, the ungraded plain, the flood plain and a number of isolated hills (Plate : 1) along the left bank of the river Son. Thus the Plateau of Rohtas includes a variety of landscape units which provide a wide and unexplored field for geomorphical study and appreciation.

1.5. OBJECTIVES

The discussion (Ch. 1.4) reveals that the Rohtas Plateau is a small museum of topographic features ranging from the tectonogenetic surfaces and forms of fluvial processes to residual and Karst landforms. Most of the features are fascinating owing to their peculiar forms and materials that render their existence. It is astonishing that many such captivating features have been left unrecorded, particularly of Karst landforms, and the literature on geomorphology, published in India, does not sufficient number of examples from India, except the one written recently by Ahmad (1985). It is for this reason that we look through foreign examples while studying geomorphology. The present study, **GEOMORPHIC EVOLUTION OF THE ROHTAS PLATEAU, BIHAR**, proposes, therefore, to make geomorphological analysis of features of the Plateau as well as to make a record of all the unexplored features to fill up the gap in Indian geomorphology.

The objectives can be achieved only by studying the area geomorphologically, because the science of **geomorphology** (from the Greek, earth or land ; **morphe**, form or shape; **logos**, science or discourse) is the study of landforms, and in particular of their nature, origin, processes of development, and material composition' (Cooke & Doornkamp, 1978, Preface). It is here not desirable to critically examine the definitions of geomorphology given by a host of scholors (viz., Lobeck, 1939, p. 3; Thornbury, 1954, p. 1; Strahler, 1960, p. 452; Challinor, 1961, p. 89; Worcester, 1965, p. 3; King, 1968, p. 403; Dury, 1970, p. 173; Sparks, 1972, pp. 1-4; Twidale, 1976, p. 3; Prasad, 1979, pp. 4-5; Kumar, 1979, p. 1; Pitty, 1982, p. 1; Faniran & Jeje, 1983, p. 1; Ahmad, 1985, preface and many others) ; because each generation of geomorphologists has redefined both the 'scales and processes in manners which have appeared to accord best with the contemporary aims of the science. The changing theoretical bases of geomorphology derive from these changing aims, which determine the character that geomorphological theories assume' (Chorley, 1978, p. 1).

The main objectives of the present book are, therefore, to make a detailed geomorphic analysis of all the geomorphological features of the Plateau's terrain on spatial and temporal scales and in terms of the processes which produced, sustain and transform them within these scales.

1.6. PREVIOUS LITERARATURE

The Rohtas Plateau forming a part of the long west-east elongated Vindhyan Range of Central India has attracted the attention of eminent earth scientists specially geologists for over a hundred years to its significance both from point of view of academic and economic interests. Therefore, many valuable references with regard to its geology and lithology are obtainable from the pioneer works of geologists like Dangerfield (1823), Williams (1952), Oldham (1859), Medlicott (1859), Medlicott (1859), Mallet (1869), Oldham, Vredenberg & Datta (1901), Vredenberg (1906), Jones (1909), Fox (1928), Auden (1933,1943), Dunn (1941), Krishnan (1943), Mishra (1948,1969), Muktinath (1948), Ahmad & Mathur (1952), Datta (1953), Choudhary (1954), Bose (1955), Howell (1956), Pascoe (1959), Krishnan & Swaminath (1959), Wadia (1961), Ahmad (1962), Awasthi (1962), Sahni (1962), Lahiri (1964), Ray, Sah, Seth & Goswami (1966), Mathur (1970,1981), Singh (1973,1980), Srivastava & Mehrotra (1975), Srivastava &Iqbaluddin (1976), Das (1977), and a host of others. These records preserve some accounts of landforms too. Moreover, geomorphology is the study of landforms, materials and processes. For **materials** these records have been treated as fundamentals, and no geomorphological research can be carried out without the knowledge of materials that make the surface.

It is however, noted that no systematic and detailed work on geomorphology of the Rohtas Plateau was done previously. Some attempts have been made by Chhibber (1948,1950,1954), Singh (1958,1969), Ahmad (1962,1976), Chakraborty (1965), Choudhary (1966), Singh and Kumar (1970), Singh (1971), Dutt (1971), Prasad (1972), Pathak and Pandey (1975), Pandey (1975-76, 1976,1978), Prasad (1983). All these works are either based merely on the interpretation of topographical sheets and geological records or they give scanty information of the features of geomorphological interests. No exhaustive work supported by intensive field study and on the basis of detailed topographical sheets prepared in recent years has so far been presented yet by any serious workers.

We can rather make a record of the novel, **Bhutnath** (the first novel in Hindi), written by Durga Prasad Khatri (1940, pp. 1-110), the then Forest Officer of the region. It appears that Durga Prasad Khatri travelled widely in the region including the intricate tunnels of caverns. Khatri, through his main character Bhutnath, has presented a description of the caverns of the Rohtas Plateau, which is proved to be true today. According to him, hidden amid the boulders was a cave. They (Bhutnath and his associates) entered into the cave and after a distance of about 25 yards, they came to a crossroad from where the tunnels were bifurcating into four directions, **i.e.**, towards left, right, ahead and the way they entered through. The tunnels running ahead and towards left and right were so narrow and congested that two persons could not go through side by side.

Almost a similar description of the caverns we have from Prabhakar Singh, another character of the novel **Bhutnath**, when he says that the tunnels were narrow, dark and intricate.

This shows that Khatri travelled widely in the region.

1.6.1. GEOLOGICAL RECORDS

Captain Dangerfield (1823) appears to have been the first scientific observer to comment on the rock types of Vindhyan system of Central India, and thereafter several casual observers noticed them in different areas. D.H. Williams, in the early part of (1848) visited the

Kaimur Range of hills, to the west of the Son river, to examine a reputed occurence of coal near Dehri. He was accompanied by J.D. Hooker, D.W. Williams (1852) of Geological Survey of India, was perhaps the first trained geologist to examine the Kaimur Mountain in detail. In 1859, Oldham also examined part of the Vindhyan formations and gave a summary of his observations to the Asiatic Society in 1856. He classified the formations into Kaimur, Rewah and Bhander, and this continous to be in use except for the fact that Semri series has been added at the bottom. Thereafter, H.B. Medlicott (1859) reported on the Vindhyan rocks of Bundelkhand, while his brother J.G. Medlicott (1859) examined those of the Narbudda and Son valleys, recording the presence, for the first time, of what he termed the 'Sub-kymores'. These have come to be known as the Semri series, or more correctly the Semri groups.

The earliest comprehansive study of the Vindhyan system as a whole was, however, brought out by mallet in 1869, who, while retaining the above mentioned three fold classification of Oldham, supported the addition of the **Sub-Kymore** underneath, re-christening them as the Lower Vindhyans. This took root immediately, and gave the Semri group a prominence quite out of proportion to its importance. This two-fold subdivision has, like a bad coin, remained in circulation in spite of the fact that it has been countered by many a field observers. Mallet, however, gave a valuable description of the stratigraphic sequence of the Vindhyan rocks from the area now reported upon, and perhaps no other published report on this area exists. This system of rocks has received attention from several geologists, notable among them is Fox (1928).

A large part of the Vindhyan basin was later examined by R.D. Oldham, Datta & Vredenberg (1901) and many of their observations are valid even today. Thus, they were the first to point out the drainage was very young in the area and that there was no noticeable unconfirmity between the so called lower and upper Vindhyans. Vredenberg (1906) has suggested the classification of the Vindhyan system. Jones (1901) has reported on the formations of Vindhyan. Fox (1928) examined the Son valley section and also opined that there was no noticeable unconfirmity between the Semri and the Kaimur groups, and hence the boundary between the lower and the upper Vindhyans should, if at all, be placed below the horizon of Susnai breccia, higher up in the succession, on the presumption that these may represent some significant disturbance in the basin. He, however, advocated a detailed re-examination of a large part of the basin in Mirzapur district and this was undertaken by Auden (1933) with results that have become house-hold with any student of Vindhyan geology. He mapped a narrow belt of the country immediately to the west of the area now taken up, although he had the limited objective of determining where the boundary between the lower and the upper Vindhyan should be placed in the stratigraphic column, and incorporated in the geological map of the country that was under preparation then. He differed from Fox about the significance of the Susani breccia and pointed out that not only there was no unconfirmity between the so-called lower and upper Vindhyans but the Susnai breccia itself had no fixed geological horizon and instead appeared to be haphazardly distributed in space and time. Auden's paper remains by far the most important work on the Vindhyans system of the Son Valley. J.A. Dunn (1941) in his book on Bihar, described the geology of Kaimur Plateau. Auden (1943) has pointed out the age of the Vindhyans. Krishnan (1943) has given a general description of the Vindhyan system. Mishra (1948) examined the evidences of the existence of life during the Vindhyan period. In

1943 Muktinath of G.S.I. re-examined the deposits within Mirzapur district with a view to locate flux grade limestone. Ahmad & Mathur (1952) described the Vindhyans in the Son valley of Mirzapur. A.K. Datta (1953) has pointed out the life in the Vindhyan period. Choudhary (1954) has examined the occurence of porcellanites in Semri series. Bose (1955) in his proper pointed out the plant life in the Vindhyans. Howell (1956) studied the nature of fossils. Pascoe (1959) gave the general information about the Vindhyan system in India. Krishnan and Swaminath (1959) have however, expressed the opinion that there was an unconfirmity between the Semri and the Kaimur groups, it bring the result of a period of earth-movement at the close of the Semri period of sedimentation. This, as indicated above, does not confirm the opinion of those who have had the apportunity to examine the formations in any detail in the field, and need not be relied upon. Wadia (1961) has written about the Vindhyan system in his book 'Geology of India'.

Recently, a number of geologists have examined the Vindhyan rocks, but their observations have been confined to restricted area or limited objectives. Ahmad (1962) has studied the palaeography of the entire Vindhyan basin in Central India, basing it mostly on the work of earlier authors. He expressed the opinion that the Son valley was a bay and it joined the main Vindhyan basin farther west. This main Vindhyan basin extended from near Agra in a south-east direction and perhaps passed out of the present land area somewhere to the south-west of the Narmada valley. Ahmad (1962) has pointed out that faulting in the Vindhyan system is very young and scarp has been formed by a more rapid erosion of the shattered rocks than those on the plateau, while Misra and Awasthi (1962) have compiled, in a very useful form all the information available about the organic emphasis on the significance of the Salt Pseudo-morphs appears to be rather misplaced. Sahni (1962) has explained the Vindhyan system of India in general. Lahiri (1964) explained the petrology of the Vindhyan rocks around Rohtasgarh. Systematic mapping and investigation was carried out in Rohtas belt by the Director of Geology, Bihar, during 1965-66. M/s. Hindustan Steel Limited and M/s. K.L.C.W. Banjari drilled a few bore-holes in Jadunathpur-Nawadih area to study the extent and quality of limestone of flux grade and cement grade respectively. Ray, Sah, Seth & Goswami (1966) have surveyed the area of Shahabad for underground water resources. Mishra (1969) in his excellent presidential address to the geology section of Indian Science Congress has ably summarised the Vindhyan rocks. In 1970 Mathur confirmed the long-held view of the basic unity of the entire Vindhyan supergroup since the Semri and Kaimur groups are perfectly conformable and there is lack of justification for division of the Vindhyan into two major, but unequal subdivisions, namely the lower and the upper Vindhyan. Singh (1973) has explained the depositional environment of the Vindhyan sediments in the Son valley. In 1975, Srivastava & Mehrotra studied the Bijaigarh shales in Central Son Valley. Srivastava and Iqbaluddin (1976) have pointed out the recent observations on Vindhyan sedimentation in the Son Valley, Mirzapur district. Das (1977) has explained the methods of detailed exploration of bedded limestone deposits, with a reference to Amjhore Pyrites. Singh (1980) has explained the Bijaigarh shale, Vindhyan system is an example of lagoonal deposits. Recently, Mathur (1981) has described the cyclic sedimentation in the Vindhyan supergroup.

Thus the Vindhyan formations in general and the Rohtas Plateau in particular have been studied not only in detail but their re-examination and reassessment has been made regularly. These continued efforts are not hypothetical but are based on intensive field study by the geologists of the Geological Survey of India, State and Central Government agencies involved in assessing the economic value and by the academicians coming from universities and various research organizations. The geological study has clearly illustrated the **materials** that make the Plateau under study. Therefore, there is probably no difficulty in analysing the landforms particularly with regard to the materials that make them.

1.6.2. GEOMORPHOLOGICAL RESEARCHERS

In spite of sufficient information about the materials of the Rohtas Plateau, no detailed work has been presented on geomorphology of the area. It has reasons. As we have noted, geomorphology is the study of landforms and their materials and processes which make them and shape them. The analysis of landforms and the processes involved requires intensive field study facilitating close observations. The analysis is done sometimes on the basis of published maps. The geomorphological study of the Plateau remained superficial because no researcher bothered to go to the interiors which are not easily accessible and have desolation, and whatever study was done on the basis of the topographical sheets, was based on old maps which showed no contours of the Plateau surface. There were some hachure-like features drawn on the old maps. This could not project the real picture of landforms. Recently topographical sheets on 1:50,000 scale have been made available showing drainage and contours in detail. But the area is inaccessible even today, having no road. The area is not facilitated even with bullock-cart. It is not developed even for bicycle-travellers.

Therefore, whatever geomorphological information we get from the published articles and unpublished theses and dissertations, is generalized and sketchy.

We have three articles from Chhibber (1948,1950,1954), which deal with the origin and nature of plateaus and drainage of India in general: but particular attention has been paid to the Rohtas Plateau. He has discussed the nature of scarp, its recession and the shifting of the Son river. His observation is remarkable when he says that the streams flowing over the plateau and scarp have been responsible for the recession of the scarp which have existed originally on the edge of the Son.

Singh (1958) observed that the Plateau forms flat tops and the steep scarps at their edge overlook the Son valley and the Shahabad Plains. The juvenile character of the kaimur Plateau edge contrasts strongly with undulating surface of its top.

Ahmad made some geomorphic observations of the Rohtas Plateau in 1962 and 1965. Following Dunn's view he supported the theory of successive uplift during the Himalayan orogeny in the Tertiary periods. So, the Plateau has gained the present height due to Tertiary uplift. 'The escarpments are the most prominent features of the Rohtas Plateau. The steepest scarps occur where the stratum is wholly sandstone. When sandstone is underlain by shale, the scarps consist of a vertical precipice above in sandstone and undercliff of about 30^0 slope below' (D.G., 1965, p. 352).

Chakraborty (1965) observed that the faster rate of valley-side widening (slope recession) than valley-bottom deepening indicates that gravity-sliding and fluvial transport are the two primary vehicles of mass-wasting. He noted six main topographic features right from the Son river to the Plateau-top.

Singh and Singh (1971, p. 646) observed, probably for the first time, that the Plateau is dissected. Similarly we have the first report of the Plateau as a whole from Dutt (1971), but it seems that his study based purely on the topographical sheets, may be an old one; because he observed that the Plateau is flat, which is not the reality. The author of the thesis studied the Kaimur Plateau, to which belongs the Rohtas Plateau, in 1972 in the form of his M.A. dissertation. He travelled the area, no doubt, but the study was based on the old topographical sheets, because he could not go into such details as it required and as he feels now. He, however, presented some lively photographs of interesting topographic features like caverns, scarps, waterfalls, residual hills, **etc**.

In the form of a Ph. D. thesis, the geomorphic analysis of the Plateau comes, for the first time, in the thesis entitled 'Geomorphic Study of the Kaimur Plateau' by J.N. Pandey in 1978. It is sorry to note that the work also is based on the old topographical sheets and has not been supported by an intensive field study. This is why the work is merely a sketchy description of general topographic features. The plateau-top has been declared flat, although it is not so.

Most of the geomorphic study of the area has, thus, remained sketchy. It is only due to the lack of intensive field study which is the general trend of geomorphological studied in India in particular. It is a fact that the area is inaccessible having nothing of comfort, rather it is full of trouble right from the travelling on foot to the terror of dacoits and carrying the essential things on shoulders whatever was required for the journey day after day. The author has not been able to project all the geomorphic details into his thesis to his full satisfaction, but he has tried his best and he has, at least, taken the trouble to traverse the area widely. He has tried to fulfil the task given by Chakraborty (1965, P.107) who says, ' the purpose of this essay is to place before more competent workers in the field of Indian geomorphology a set of observations made during the said tour as postulates on the mechanism of landscape evolution on a small part of the Vindhyan country'.

1.7. SOURCES OF INFORMATION

Geomorphology has already been defined as the study of landforms, their materials that make them and the processes which shape them. The geomorphological study of the Rohtas Plateau needs, therefore, a co-ordination of all these aspects. It is also a fact that for an individual it is rather impracticable to hunt out each and every aspect of the study on his own. Therefore, it is feasible to collect information from different sources and make use of them in his geomorphic analysis. However, a critical examination of the information is essential while analysing the landscape.

We have pointed out (Ch. 1.6) that there is an ample record of the materials with regard to the Rohtas Plateau. These records have been beneficial in analysing the stratigraphy, geology and structural characteristics of the area in detail.

Before going to the field, the topographical sheets, both old and new, published in 1955 and 1971 respectively, have been useful in giving a general idea of the surface drainage and landforms of the area. The area is covered by the topo-sheets numbering 63 P/5,6,9,10,13,14, and 72 D/1,2 on the scale of 1:50,000. For detailed study of the area the topo-sheet on the scale of 1:50,000 have been preferred.

Many topographic features which were not clear from the topo-sheets were verified by the author in the field. He travelled the area many times right from his childhood to the occasion when he prepared his M.A. dissertation. But for the purpose of the present work, in particular, he along with his supervisor Dr. Nageshwar Prasad and five other friends traversed the whole Plateau from south to north, **i.e.**, from Baulia to Karamchat Dam area **via** Rohtas Fort, Budhua Kala, Luka, Bunda and Gupta Dham in February, 1984 on the occasion of Shivaratri. The occasion was chosen because a large number of devotees visit the Gupteshwar Dham to offer their Puja to the Lord Shiva. This time is free from terror and desolation. Several times he visited the different limestone quarries and pyrite mines. In 1985, in the month of November the author travelled in the Son valley from Basuhari (south-western portion of the Plateau) to Japla Cement Factory (south-eastern portion).

The journey has unfolded many details of geology and structure of the area. Although the reports and memoirs of the G.S.I. have information about the geology and structure of the Plateau, some new and marked characteristics have been observed. This will be clear from the subsequent chapters. The books by Pascoe (1959), Dunn (1941), Wadia (1961), Krishnan (1982), and other eminent geologists have been very helpful. Rock samples have revealed the structural characteristics of the rest of the area. The extent, grade and expected reserves of the limestone deposits in the Plateau have been given by various agencies like the Indian Bureau of Mines; Geology and Mines Department, Govt. of Bihar; Kalyanpur Lime and Cement Works,Banjari,Son Valley Port land, Pyrite, Phosphates & Chemical Limited, Amjhore. The author has collected the limestone from quarry, dripstone and dripwater from caves and got the water and stones analysed in the Chemistry Department of B.S.K. College, Maithon.

Geological maps have been obtained from Geologists and Surveyors of different agencies. From these maps faulting and folding in the area have been confirmed.

Many photographs taken from the field have helped in the analysis of landforms of the Plateau.

Many books and journals, both from India and abroad have helped much in the literary expression and description of the landforms in the Rohtas Plateau. These are listed at the end of the present research work.

As regards climatic data, the author obtained them from different ***anchals*** of the Rohtas district and the Statistical Office, Sasaram.

Since natural vegetation is considered to be one of the chief agents in shaping the geomorphic features on the face of the earth, data and maps concerning distribution of natural vegetation and types of trees have been obtained from the office of the District Forest Officer, Sasaram.

Recently, in 1980, a detailed report of limestone reserves found in Jadunathpur, Nawadih and adjoining areas in Rohtas District, Bihar has been prepared by the Directorate of Geology, Department of Mines and Geology, Govt. of Bihar. This has revealed the structural characteristics of the area.

Last, but not least, Dr. Prasad's suggestions and information are actually the fundamental source to the author and he is indebted to Dr, Prasad.

1.8. METHODS OF GEOMORPHOLOGICAL ANALYSIS

It is worthwhile to consider the various methods and approaches to the geomorphology of the Rohtas Plateau. Geomorphology is essentially a field study. The emphasis has, therefore, been laid, wherever possible, on the field work. It is also an accepted fact that all the problems in the field cannot be solved mainly because of the great complexity of geomorphological processes in nature. However, 'field studies play a vital part in nearly all geomorphological investigations' (King, 1966, p. 31). Observations in the field are essential either to formulate or to test hypotheses and theoretical calculations, and they must also accompany model work. The author was, therefore, obliged to make journey to the field for his observations, as stated earlier. This study involved three principal processes-observation, recording and interpretation. The author had taken all pertinent materials-maps and statistics, **etc**. to the field.

The first item in his equipment was the map records of all the work previously done in the area. The topographical sheets on 1:50,000 scale were the most useful.

But this work would be more minute and detailed than if it is, if the author had various modern equipments. Unfortunately no such instrument was available to him. He might be able to measure the frequency, rate and movement of slopes, rate of soil erosion or mass movement, **etc**.

However, much more could be observed concerning the landform of the region by simple methods, as suggested by Dr. Prasad, under whose kind supervision this work was completed. The author has noted the shape of scarps, plateaus, hillocks and valleys, roughly the degree and frequency of slopes, the development of drainage pattern and the nature of the material exposed in all available sections. He noted the dip of rocks of different areas, particularly in the quarries by simple equipment, e.g. goniometer. This gave a rough, but reasonable idea of the region.

It is a peculiarity of the Plateau that it is devoid of any means of communication and transport. Not even a single road can be developed through the steep and rocky face of scarps. The only road which runs through the notches on the extreme north-west connects the Adhaura

Block head-quarters to Bhagwanpur. Therefore, whatever field study has been done on the top of the Plateau, is absolutely based on the journey on foot. In the foot-hills region, however, he travelled by bus and preferred halting at various places around which the characteristics of the area could be observed.

Besides these maps and notes, the work is supplemented by photographs illustrating specific landscapes or important details. It is often said 'any fool can use a camera'; but 'it is a particularly foolish geographer who neglects to take plenty of illustrative photographs' (Putnam, 1962, p. 403). So, the author took as many pictures as possible, either it was the common and more or less monotonous feature of the area under study or the most striking feature as a whole, for he might not visit the place again. It has become easy for the author to discard unrequired photographs, while an apportunity not utilised to the full might result in lasting regret.

On the basis of various maps previously prepared and the field study done by the author, the geomorphology of the Rohtas Plateau could be described. Geomorphology is basically the study of contours. Map contours tell much more than the story of relief, the differences in height above sea-level. 'to the trained eye of the geographer and geologist they also tell the story of the way in which the relief was developed' (Putnam, 1962, p. 407). But for the author it required some laboratory studies. He drew many long, cross and valley-way profiles within short intervals and then the area was interpreted easily. For slope analysis of such an area, he preferred wentworth and Raisz & Henry methods. Graphical method for reduction and enlargement of maps have been suitably applied. The area was measured by graphical methods and also by planimeter. Rotameter was used in measuring the length of streams. Many methods have been dealt with within the topics. In brief, various methods stated on maps and descriptions follow the book of Monkhouse (1970), although the books of A.K. Lobeck (1939), A.M. King (1966), King & Doornkamp (1971), and G.H. Dury (1970) and many others have been consulted. The author has drawn maps, diagrams and profiles numbering more than 150 to enrich his thesis and to make the interpretation of the geomorphic features easy and comprehensible.

Many maps related to the texture of drainage, hydrographic network and relief measures, etc. have been prepared to study the drainage and morphology of the Plateau. The topographical sheets of 1:50,000 have been preferred therefor. All the values have been counted in one square kilometre grids of the entire area and their respective figures have been inserted on a 1:100000 map to prepare maps by isopleth method. Detailed study of the Rohtas Plateau has been included in this research work, mostly by maps and field study.

So far as the geology and structure of the area is concerned, much information is based on the previous work, and they are examined and verified by rock samples taken from the field, and identified and interpreted in the laboratory of Japla Cement Factory, Palamu, and the Chemistry Department of B.S.K. College, Maithon.

'Like many other branches, geomorphology is becoming more quantitive in its method, and statistical techniques play their part in many stages of study' (King, 1966, preface). In the present work the approach is non-mathematical for the most part, but makes use of elementary principles of physics and chemistry in explaining the action of natural processes. Simple mathematics, however, has not been neglected.

As regards the description of the region, the analytical method has been adopted with justification. This system attempts to present a simple and reasonable explanation of each object or form it describes.

At last, it may be said that most of the methods, applied to the region, are simple, less complicated. Many methods, the author feels, are based upon complex equipment which is beyond the scope of many geomorphologists in India; and as such those methods could not be adopted to the study of the Rohtas Plateau.

2

MATERIALS AND THEIR STRUCTURAL CHARACTARISTICS

Rock samples collected from different parts of the Rohtas Plateau, observations in the field, discussion with the geologists directly associated with the area particularly with the limestone quarries and study of numerous books, articles and maps including geological records, memoirs and bulletins published by the Geological Survey of India and available with the private and government organizations of cement factories and development authorities reveal that the materials that make the Plateau under study and which play a dominant role in its geomorphic evolution consist predominantly of limestone, shale and sandstone of the Vindhyan periods. On the south, north and east it comprises of the alluvium which stretches in narrow elongated belts and which are recent and ultra-recent origin. The materials of the Rohtas Plateau are structurally folded, faulted, intruded, warped and tilted.

The study of materials and their structural characteristics has been necessitated in the context of the geomorphic evolution of the Rohtas Plateau because these are the fundamental elements which play important role in the origin and development of the landforms. The discussion will go through all those ways in which 'the earth materials out of which landforms are carved, differ from one another in their physical and chemical attributes. It includes such phenomena as rock attitudes; the presence and absence of joints, bedding planes, faults, and folds; rock successiveness; the physical hardness of the constituent minerals; the susceptibility of mineral constituents to chemical alternation; the permeability or impermeability of rocks, and various other ways by which the rocks of the earth's crust differ from one another' (Thornbury, 1954, p. 17).

The Rohtas Plateau comprises rocks of the Vindhyan formation mainly limestone, shale and sandstone. The sedimentaries have been subdivided into two subdivisions, the lower Vindhyan and upper Vindhyans. The lower Vindhyan consist of marine calc-argillaceous sediments whereas the upper Vindhyans characterise almost wholly of red coloured sandstones and subordinates shale laid down in lacustrine and fluviatile environment. The plane separating the two is normally a well marked disconfirmity. The upper Vindhyan sediments cover four wider area than the lower Vindhyans. There is an unmistakable evidence of transgression overlapping the red non marine sediments of the upper division on the marine sediments of the lower Vindhyans.

The upper Vindhyan sediments are three to four times thick as the lower Vindhyans. Thus the great difference in the lithological associations, petrological attributes and sedimentary structure amply justify the subdivision of the Vindhyans of the Rohtas Plateau into two parts.

Besides the Vindhyan formation, the study area consists of materials originated due to weathering, mass wasting and other denudational activities over the fundamental rocks through the ages. The talus which was transported from the scarp face and the Plateau top forms the looser foundation of the flood plain where rock fragments of varying sizes are found buried under a thick mantle of river-borne materials. The coarse materials received from the Plateau and the fine materials transported with the Son water particularly during the floods produce alternating layers of coarse and fine materials in the flood plain. A brief discussion of these materials have been made under the heading 'soil Stratigraphy'.

Here it may be mentioned that the Vindhyan rocks comprise almost 88.64% of the study area, so it is natural that emphasis has been given on the discussion of the materials belonging to the Vindhyans. The materials and their structural characteristics of the Rohtas Plateau may, therefore, be discussed in detail under the following heads :

2.1 AGE OF VINDHYANS

Although the Vindhyans have been regarded to be the Pre-Cambrian formations, the last of the oldest sedimentary strata by many of the geologists (Mishra & Bhatnagar, 1950, pp. 88-89). The age of the exposed Vindhyan sediments has been a controversial topic. Geologists and palynologists in India have given different opinions in this regard.

Oldham (1893) suggested Cambrian age for the Vindhyan Supergroup. On the other hand, Auden (1933) stated that the age of the Vindhyan rocks may be any thing from Algonkian to Devonian. Rode (1946) indicated early Palaeozoic age for the Rohtas limestone. Based on a primitive form of dasycladaceous alga, Mishra (1949) suggested Devonian age for these sediments. Sitholey (1953 et al.) assigned Cambrian age to the upper Vindhyan beds. Sahni and Srivastava (1954) described fossils from the Suket shale and assigned to them Cambrian age. In 1956, Howell discovered some disc-like fossils comparable to Morania of Cambrian age. Bose (1956) found that the winged spores of Archegoniate group indicate Lower Cambrian age for the Basal and Porcellanite formations of the Semri group; middle part of the Upper Cambrian age for Kheinjua and Rohtas formations (Semri group); upper Cambrian age for the Kaimur. According to Pascoe (1959), 'all that is definitely known concerning the age of the Vindhyan system is that it is older than Gondwana and younger than the Gwalior (Dharwar)'. Krishnan (1968) believes that the Vindhyan may be mainly of Cambrian age. Maithy (1968), on the basis of acritarchs and other algal remains from the Suket shale of Rampura (M.P.) concluded that the entire Vindhyan supergroup is not Precambrian in age as was thought earlier and that at least some of the strata are definitely younger in age. Pichamuthu (1971) has indicated that the base of the Vindhyan is probably 1400 million years and the upper Kaimur about 910 million years. Shrivastava (1975 et, al.) assigned Ordovician to Silurian age to the lower Vindhyan. Venkatachala and Rawat (1972), reviewing the earlier data on the Vindhyan sediments, have suggested Precambrian to Cambrian age. Maithy and Shukla (1977) have assigned late Pre Cambrian age to the Suket shales of Rampura (M.P.).

From the above discussion it can be concluded that the age of the whole of the Vindhyan sequence ranges from post-Cuddapah to late Pre-Cambrian. The author has, therefore, tried to trace out of the surface history (Ch. 3.1.2) right from the Pre-Cambrian to the Tertiary.

TABLE : 2.1

Lithostratigraphic Subdivision of the Vindhyan Supergroup (Modified after Auden,1933)

Group	Formation	Member
Upper Vindhyan Group	Kaimur formation	Dhandraul quartzite
		Scarp sandstone
		Bijaigarh shale
		Upper quartzite
		susnai Breccia
		Silicified shale
		Lower quartzite
--------------	Unconformity	--------------------
Semri Group	Rohtas formation	Limestone & shales
		Glauconitic sand-stones
	Kheinjua formation	Fawn limestone
		Olive shales
	Procellanite formation	Porcellanites
	Basal formation	Kajrahat limestone
		Basal conglomerate
----------------	Unconformity	---------------------

After Kumar, 1982, p. 103

2.2 LITHOLOGY

The general lithostratigraphic succession of the rocks of the area is given below :

2.2.1 SEMRI SERIES (Lower Vindhyan)

The term Lower Vindhyan as best can be informal. It merely represents a group of rocks that form distinctive lithostratigraphic succession, with carbonate rocks, tuffs and acid volcanics forming the dominian to components in the Son valley. Sandstone forms only a minor constituent. Interpreting basin tectonics from sedimentary characteristics, Prakash (1966) and Banerjee (1964) have emphasized the fact that the Lower Vindhyan (Semri) was formed in isolated, negative basins. And this was followed by development of the vast traansgressing Upper Vindhyan sea.

PLATE : 1. STRUCTURAL CHARACTERISTICS

Murli Hill separated from the main Plateau by the Ausane Valley near Banjari showing horizontal beds of the Vindhyan sandstone (A) and the limestone bed (B) very conspicuously. On the east lies the flat agricultural alluvial tract (C). It is a clear view of sandstone.

Plate - 1

TABLE :2.2

THE SEMRI SERIES (EASTERN BASIN)

Rohtasgarh (Lahiri, 1964)	Southern Mirzapur (Auden, 1933)
Baulia limestone Grey limestone, yellow and red shales	Rohtas stage Marine grey limestone with nudules and interbedded calc shales
Amarkha sandstone (green) with shales	Kheinjua stage Glauconite sandstone Fawn limestone Olive shales Porcellanite stage Basal stage Kajrahat limestone and conglomerates

After Mishra, 1969,p. 115.

The Semri series comprises predominantly limestones and shales with Glauconitic sandstone. Some of the limestone are highly magnesian. This series has four stages (Table : 2.1), but however, in the study area is exposed the basal stage, although it is certainly overlain by the porcellanite stage. The author considers it logical, therefore, that the lithology of the basal stage not be discussed in detail, rather a brief outline is justified because the limestone portion below can affect the landforms.

2.2.1.1 BASAL STAGE

2.2.1.1.1 Basal conglomerates

This consists of a variable series of pudding stones, conglomerates, breceias, brown quartzites, bleaching shales and lenticular limestones. It is inconstant, and the failure to crop out cannot always be assigned to faulting, but must be regarded as due to original capricious sedimentation. It grades into pebbly and gritty quartzites and finally to quartzites without pebbles.

The underlying Bijawars are here purple slates and phyllites, without jasper, so that it is necessary to assume the rolling down of scree from an area of jasper rocks, once at a greater height, and long since denuded.Strong unconfirmity between basal rocks and underlying Bijawar is found.

2.2.1.1.2 Kajrahat limestone

This limestone varies considerably in appearance. It is sometimes coarsely crystalline and grey blue in colour, but more usually it is finely to microcrystalline and of blue, brown and white tints. It is well bedded, occurs in layers of from four inches to two feet thick, and is marked out by the finer bands of varying hardness which are with difficulty recognised on a fresh fracture surface. The more outstanding of these are siliceous and non-reactive to dilute hydrochloric acid.

The rock of basal stage cropped out in and around Bhawanathpur area in the district of Palamu, which is about 10km. south of the Rohtas Plateau.

2.2.1.2 PORCELLANITE STAGE

The porcellanite stage, consists of silicified or porcellanised tuffs, shales, sandstones. The predominant and typical member of this stage is a hard, fine-grained, porcellanic rock, varying in colour from almost black to a pale green, and braking with a conchoidal fracture into extremely sharp-edged splinters. 'The most interesting are some ancient volcanic beds remarkable for their porcellanic texture' (D.G., 1966, p. 17). In the Son Valley the rocks of porcellanite stage have been found near Kadhwan.

The porcellanite stage represents an original marine formation deposited within the vicinity of volcanoes. The exposures of the Porcellanite stage in the outliers forming the centre of the Son synclinorium seem to be less in thickness and on the average finer in grain than they are along the river bed is to a certain extent, an indication that the volcanic centres lay to the north of the Son.

2.2.1.3 KHEINJUA STAGE

The Kheinjua stage is named after some hills south of the Kaimur scarp at and beyond the western boundary of Rewah State. Its rocks are principally shales and sandstones but limestones are occassionaly found. South-west of Rohtasgarh in the lowground intervening between the foot of the scarp and the bank of the Son Kheinjua limestones are seen for a short distance. It is absent in the south-eastern and eastern part of the scarp.

The Kheinjua stage consists of the following rock-formations:

2.2.1.3.1 OLIVE SHALES

Following above the porcellanite are fine grained olive-green shales with cuboidal or pencil fracture over the whole of the area. There is no actual junction seen between these shales and the Porcellanite stage.

2.2.1.3.2 FAWN LIMESTONE

This is a fawn-weathering, siliceous and cherty limestone, with an interior which is seldom clear, but usually shows sign of dolomitisation, silification, amygdales of agate, and segregation of lime material as calcite and veinlets. Chert is very common. Sometimes there is separation into fine grained, flinty limestone and coarser limestone, the former occuring as faulted and folded plates in the latter. Calcium, magnesium and ferrous carbonates are all present.

2.2.1.3.3 GLAUCONITIC SANDSTONE

This is a variable series of silty sandstones with the general presence of glauconite, of calcareous glauconitic sandstones, quarzites, conglomerates, breccias, and then thin limestones. The limestone bands occur near the base, just above the main Fawn and are interbedded with current-bedded quartzites.

Bedding is markedly lenticular, and seldom over six inches thick in the silty sandstones, but reaches six feet in the quartzites and conglomerates. The whole series is abundantly ripple-marked and sun-cracked; proof of shallow water conditions.

The characteristics features of this group are the presence of glauconite and the abundance of ripple marks and sun cracks.

2.2.1.4 ROHTAS STAGE

This is the top stage of the Semri series, and named after the rocks exposed below the Kaimurs at Rohtasgarh.

The uppermost beds of kheinjua stage pass up through calc-nodular shales into the most typical member of the succeeding Rohtas stage, which is a thin-bedded, flaggy limestone. The fine, parallel, grey and black banding of the shales passes also through the limestone nodules, Which vary from six inches to three feet in diameter.

The chief feature of this stage is the limestone, which is 700 feet thick. An analysis of this rock shows 84 per cent of CaCo3, less than 3 per cent of Mgco3, and something like 10 per cent of Sio2 ; the stone is used for cement manufacture as well as for the lime-burning and building stone. It is in this limestone that sacred **'Guptadham'** cavern exists. The bedding in both the limestone and shales of this stage is generally thin and markedly parallel, in this contrasting with the lenticular bedding of most of the beds in the Kheinjua stage and with the thicker bedding of the Kheinjua limestones. Typically, the bedding of Rohtas limestone is from three inches to a foot thick. It has been suggested that the conditions of deposition of the Rohtas limestone is laid down on very shallow marine flats of wide extent. Towards the top of the stage, silicification of the limestone is sometimes strongly pronounced. Near Rohtas, in the middle of the stage, is a thin bedded limestone with small scattered crystals of blue and purple flour-spar.

The charactaristics and the distribution of this important and relatively more apparent stage discussed so far can be noted as follows :

i) 'Rohtas limestones form the undercliff facing the Son along the southern and south-eastern edges of the Kaimur out-crops, and are also seen, further north, in the deep river gorges like those of the Durgawati, where they are cut through the anticlinal axes of the shallow corrugations which, striking a little north of east, help to diversify the surface of the eastern termination of the great Kaimur Plateau' (D.G., 1966, p, 16).

ii) Shales, slates and limestones met with at Chenari, Mohania, Bichhia, and Ramgarh are likely to correspond to parts of the Rohtas stage of Semri series (Ray, et. al, 1966, p. 14).

iii) Map shows that rocks of Rohtas stage are also visible in the northern scarps face in the valleys of Dumarkhar, Durgawati, Karsotha, Sura, and Karamnasha. Gupteshwar cave is one of the best examples.

iv) There are two main groups or horizons of limestone in the Rohtas formation of Shahabad. The upper limestone beds are found above the characteristic nodular limestone bed and are exposed along the base of the Kaimur Plateau in the eastern part of the belt from Chutia to Ramdihra. There are three main limestone bands in this upper group.

v) Regionally, the limestone of Rohtas stage can be described into two broad sub-divisions viz. the upper limestone presently catering to the needs of two cement plants located at Japla and Banjari and the lower limestone being quarried at Pipardih and feeding mostly the two cement plants located at Dalmianagar. The nodular limestone which serves as a typical marker horizon together with 100 m. thick overlying shales divides these two limestone horizons.

Structurally, the area, in general, is undisturbed with the little or no change in attitudes of the beds. The limestone bands show general E - W strike with northerly dip amounting 10^0-12^0. The beds show a general downwarp swing to the north and consequently the dip became steeper up to 25^0 near their contact with nodular limestone.

2.2.2 KAIMUR SERIES (Upper Vindhyan)

The Upper Vindhyan is a dominantly arenaceous facies transgressively overlying the Lower Vindhyan. A vast erosional unconfirmity separates the Rohtas limestone, from the Kaimur sand-stones, the Kaimur resting over tilted, deformed and eroded Rohtas (Mishra, 1969). 'The Kaimur sandstone occupies the greatest superficies. It covers the greater portion of table-land and along its southern and south-western border forms the upper portion of the precipitous escarpments that overlook the Son, while along northern edge it constitutes the whole height of the cliff down to their base where they meet the Ganges alluvium' (D.G. 1966, p. 16).

2.2.2.1 LOWER KAIMUR SANDSTONE STAGE

This stage consists of well defined bands of quartzite, cropping out in two distinct scarps with an intervening band of shales, containing the Susnai breccia.

TABLE : 2.3

KAIMUR SERIES
(Eastern Basin)

Rohtasgarh (Lahiri, 1964)	Mirzapur (Auden, 1933)
Murli sandstones	Dhandraul quartzite scarp sandstone and conglomerates
Anjhore siltstone (grey, pyritic shales and siltstone)	Bijaigarh shales
Akbarpur sandstone orthoquartzite and lithic sandstone	Upper quartzites, Susnai Breccia Silcified shales, Lower quartzites

After Mishra, 1969, p. 120.

2.2.2.1.1 LOWER QUARTZITE

This is generally more massive than upper quartzite, occuring in beds up to 15 feet thick. Small pebbles occur in lenticular patches in the quartzite matrix. Current bedding is locally common. The colour is white or grey.

2.2.2.1.2 SILICIFIED SHALES

Below the Susnai breccia, and above the shallow-water top facies of the lower quartzite, there are black micaceous shales with papery lamination, together with jet black, strongly jointed, silicified, carbonaceous shales readily bleach.

2.2.2.1.3 UPPER QUARTZITE

This forms, by its dip slopes, the excellent rim which surrounds the main cliffs of upper Kaimur sandstones. The scarpslope, forming a 40 to 50 feet cliff, is everywhere conspicous, and easy to map.

The rock is generally a fine grained quartzite which has been strongly silicified. The bedding is seldom above three feet in thickness, and is generally markedly lenticular, as in the case of the lower of the quartzites. Ripple are everywhere abundant.

2.2.2.1.4 BIJAIGARH SHALES

The quartzites, micaceous sandstones and shales pass up gradually to the true Bijaigarh shales which are black, carbonaceous, fine-grained, micaceous and frequently pyritiferous. They contain occcassional ironstone bands. The upper ten feet of the Bijaigarh shales are recognised almost everywhere by bleached red and yellow colour, giving the impression of baking. These shales were probably more pyritic and owe their present colour to the oxidation of pyrites and the formation of ferric oxides and sulphates. The Bijaigarh shales, about 150 feet thick, are intensely brittle and splintery, and are frequently so black in colour as to be easily mistaken for coal. They are often impregnated with iron pyrite and have been used to a small extent to manufacture sulphate of iron. In the Durgawati valley and other northern inliers, as also south-east of Khadar, the Bijaigarh shales rest directly upon the Rohtas limestone, but along the southern scarp there intervenes a subsidiary sandstone and shale known as the lower Kaimur sandstone and Rohtas shale.

2.2.2.2 UPPER KAIMUR SANDSTONE STAGE

Above the Bijaigarh shales is a series of thin-bedded flags of predominant green colour and very often ripple-marked. The beds comprise green sandy silts, micaceous green-brown sand shales, and contemporaneous conglomerates, with discs of green shale set in green-brown shale matrix. Bedding is seldom over six inches and jointing is frequently in broad curves. These beds do not form cliffs but crop out in the 30^0 slopes at the foot of the main scarp.

2.2.2.2.1 DHANDRAUL QUARTZITES

The topmost bed of the Kaimur stage is Dhandraul quartzites. The rock is generally a fine grained white quartzite, with grain size up to 0.5 mm. and far poorer in interstial clay paste than the scarp sandstones. The colour in the field varies from pure white to white stained with purple. The grains are sometimes rounded.

2.3 ENVIRONMENT OF DEPOSITION

The environment under which deposition of the upper and lower Vindhyan took place in Rohtas Plateau differs considerably from the Semri series to others. The environments of deposition for the different lithounits, though vary from beach to barrier bar or shoal through tidal flat and lagoon, the spectrum remained invariably within shallow marine domain and the facies are found to be repeated vertically. This signifies repeated shifting of shoreline through the whole span of the Vindhyan sedimentation.

2.3.1 ENVIRONMENT OF SEMRI GROUP

The Semri Group represents mainly the products of deposition in a lagoon-tidal flat complex (Table : 2.4). The deposits in the lagoon are invariably fine-grained clastics. Parallel lamination seems to be the rule, wave ripples may develop in the siltier portions. Tidal sediments are characterized by features related with alterations of currents and slack water, intermittent subaerail exposures indicated by disiccation cracks, rain prints.

A significant feature of the Semri Group, is the presence of pyroclastic rock (porcellanite). 'The pyroclastic origin is betrayed by the presence of abundant collapsed pumice, breccia, delicate glass shards, rhyolitic fragments and deeply embayed quartz grains'. (Ghosh, 1971, pp. 154-64). In its type area in Son valley, 'it comprises agglomerates, bedded chert, banded porcellanite and conglomerate' (Srivastava, 1977, pp. 45-51). Banded porcellanite shows effects of soft sediment deforation. The banded porcellanite consists of laterally persistent coloured bands, and shows cross-lamination and graded bedding. While Ghosh (1971) attributes extensive lateral continuity of these beds to operation of turbidity current.

Table : 2.4

Lithostratigraphy and Depostional Environments of Vindhyan Supergroup (Eastern Part)

Group	Formation	Depositional Environment Singh (1973)	Depositional Environment Banerjee (1973)
Kaimur Group	c. Kaimur sandstone (scarp sandstone & Dhandraul quartzite)	Shoal complex	Barrier-beach dune wash over flat & tidal flat
	b. Bijaigarh Shale	Lagoon	Lagoon
	a. Upper & Lower quartzite	Tidal flat	Barrier bar to tidal flat
Semri Group	l. Rohtas Shale	Carbonate lagoon	--------------------
	k. Rohtas Limestone	Carbonate tidal flat	Lagoon
	j. Red Sandstone	--------------------	Tidal flat
	i. Glauconite sandstone	Tidal flat	Longshore bar Tidal flat
	h. Fawn Limestone	Carbonate tidal flat	Carbonate tidal flat
	g. Olive Shale	Lagoon	Lagoon to tidal flat
	f. Procellanite	Lagoon	Lagoon
	e. Kajrahat limestone	Carbonate tidal flat	Lagoon (tidal flat towards the top)
	d. Middle Shale (Basal shale)	Lagoon	Shallow marginal lagoon
	c. Basal sandstone	Tidal flat	-----------------------
	b. Basal Limestone	Carbonate tidal flat	-----------------------
	a. Basal conglomerate and sandstone	High gradient coastal rivers	Mainland beach

After Chanda & Bhattacharya, 1982, p.90.

The origin of the conglomerate that separate the Vindhyan Supergroup from the Pre-Vindhyan basement has been a big problem. The conglomerate occurs continously all along the southern margin in the Son Valley. A number of Geologists have attributed glacial origin to the basal conglomerates. Mishra (1969) and Ghosh (1971), however, do not consider the evidence convincing enough and regard it as a shoreline accumulation, marking the Vindhyan strandline. Their conclusion is apparently influenced by (a) relation with the inferred paleoshoreline, (b) gradual transition of the conglomerate to lagoonal carbonate (Kajrahat limestone). On the other hand, 'conglomerate as a deposit formed from coastal rivers of high gradient' (Singh, 1973, pp. 140-52). 'Textural and mineralogical maturity of the conglomerates as well as its association with quartzarenite, however, suggests through reworking of the sediments - a feature unlikely to develop in coastal alluvial fan' (Ghosh, 1971, pp. 154-64). This, of course, does not preclude the intial supply of the sediments through high-gradient coastal rivers. It can be imagined that a youthful relief existed in the marginal Pre-Vindhyan areas. The Semri Group, as a stated earlier, is predominantly carbonate, and the sandstones in it are quartzarenitic in composition, both of which point to subdued topography in the source area. 'The Semri rocks were largely a marine formation. The marine nature of the limestone is one of fair certainty, on account of their thickness and excellent bedding' (Auden, 1933, p. 216). The parallel nature of the bedding of the main limestone indicates quiet deposition away from the influence of current and wave action, and from detritals. That the seas were shallow, certainly in the case of the Kheinjua and Rohtas limestone, is suggested by the fact that associated with these limestone are the shallow water and subaerially exposed beds of the Kheinjua and Lower Kaimur stages. 'To assume great-depth for the formation of the limestone would involve too rapid variations in the level of the sea floor, between the periods of limestone and of sandstone deposition' (Auden, 1933, p. 216).

2.3.2 ENVIRONMENT OF KAIMUR GROUP

The most striking features of the Kaimur Group rocks are their mainly arenaceous facies, the commonness throughout of current bedding, the abundance at certain horizons of ripple marks, the red colour of the scarp sandstones (Table ; 2.5). 'The conditions under which the Kaimur rocks were deposited may with fair certainty be considered to have been fluviatile' (Auden, 1933, p. 216).

TABLE : 2.5

Environment of Deposition of the Upper Vindhyans

Group	Environment	Author (S)
Kaimur	High energy littoral environment with shifting current and waves	Awasthi, 1964
	Dhandrauls indicate beach environment and fluviodeltaic environments for whole Kaimurs	Auden, 1933
	Multimodal orientation of cross stratification dip azimuth imply formation on sandy intertidal flats	..ivastva and Bharadwaj, 1970
	Shallow marine to beach type of environment	Chakraborti, 1963
	Deposition in the upper portion of the neritic zone and in and above the littoral zone	Garalapuri, 1969
	High energy beach environment (deltanic or tidal flat)	Kedar Narain, 1970

After Lakshmanan, 1981, p.128

Broadly speaking, the Kaimur Group in this region comprises of two sandstone units, 'separated by the well known pyrite bearing Bijaigarh shale which shows westward transition to non-pyritiferous greenish clayey siltstone. Whereas the Bijaigarh shales has intercalations of sandy intertidalites towards its base and top, its middle part is essentially black shale' (Chanda & Bhattacharya, 1982, p. 94). The Bijaigarh, in its turn, grades upward into sandstone (scarp sandstone) characterized by large-scale cross bedding.

Recently Singh (1980, pp. 83-103) suggested that Bijaigarh shale was deposited in a coastal lagoon behind a barrier complex. According to him, its middle part represents deposits of lagoonal pond under euxinic condition : the upper and lower part being high energy deposits formed on the fringes of the lagoon. The Bijaigarh shale is developed only in the eastern part, and towards west the Kaimur Group comprises almost exclusively of sandstones.

2.4 THICKNESS OF THE DEPOSITS

The study of palaeocurrent reveals the nature and depth of the basin over sedimentation took place. In case of the Rohtas Plateau we find that the thickness of the strata increases towards north and east. 'The Vindhyan palaeocurrent has been mostly determined from cross-bedding and ripple mark in the sandstones and grain-bearing limestones' (Chanda & Bhattacharya, 1982, p. 97). The table 2.6 shows the cross bedding and current ripple.

TABLE : 2.6

PALAEOCURRENT DATA AT THE TIME OF VINDHYAN PERIOD

Group	Current direction	Area	Author (s)
Kaimur	Cross beds N. 36° W. Ripple marks N.W. parting lineation N. 25° W.	Chunar	Jafar (1966)
	Ripples marks S. 48° W. - S.W.	Rohtasgarh	Lahiri (1964)
	Cross bedding S.W.	Rewa, Gobindgarh	Banerjee & Gupta (1963)
	N. 53° W.	Jigna area	Gupta (1963)
	N. 15° W.	Badanpur Maihar area	Gupta (1963)
Semri	Ripple marks S.S. 48° S.W.	Rohtasgarh	Lahiri (1964)
	Ripple marks N.W.N.	Ghurma Son Valley	Auden (1933)
	Cross bedding N.	Kajrahat	Auden (1933)
	Cross bedding N.W.	East of Chopan	Auden (1933)
	Ripple marks W. Cross bedding N. 26° W.	Rewa Gobindgarh Maihar area	Banerjee & Gupta (1963)

After Mishra, 1969, p.130

'During Semri times in the extreme eastern edge of the basin near Rohtas the current direction is S.S.W to S. 40° W.' (Lahiri, 1964, p. 270). In the Son Valley of Mirzapur the basal cross bedded sandstones suggest westerly to north westerly currents.

Kaimur palaeogeographic of Rohtasgarh remains unchanged, mainly S.S.W. to S. 40° W., but in Dhandraul times the currents were directed northward of N.N.W. at Chunar (Jafar, Akhatar, & Srivastava, 1966, p. 82).

The palaeocurrent data obtained by Jafar (1966) from the Kaimur at Chunar (N.W. or N.N.W.) is extremely significant. This area represent the northern border of what is now called the Vindhyan basin. Jafar (1966) has been pointed out that the current directions show that the deeper part of the basin should exist north of Chunar. In other words, the original basin extended far to the north below the alluvium of Ganga, Yamuna. Jafar (1966) maintains that lower Vindhyan sea did not extend much beyond the present northern limit. But in the following Kaimur times the sea extended far beyond, submerging even the Bundelkhand massif. This northward transgression of the sea was probably accompanied by regression of the southern shoreline. Thus the maximum thickness of the deposits in the Rohtas Plateau, i.e. of Vindhyan supergroup has been estimated to be of the order of 4,500 metres. The Upper Vindhyan sequence is three to four times as thick as the Lower Vindhyan Sequence. In general, the maximum thickness of the Lower Vindhyan occurs along the southern border and gradually diminishes towards the north and northwestern part of the basin (Banerjee, 1964).

Palaeogeographic map drawan by Ahmad (1958) for the Semri period shows its maximum development near Chopan (24° 32' N. and 83 02' E.) in the eastern part of the Son Valley. The thickness of the Semri Group at Chopan, as measured by Auden (1933) is about 1120 m. Towards west, the basin becomes shallow, wide and open, and the semri Group attains constant thickness (~304 m.). Ahmad in his isopach map has shown maximum thickness of the Kaimur Group at place north of Rohtasgarh (24° 30' N. & 83° 40' E.) and the basin continued to remain open, wide and shallow towards west during the Kalimur times as well. Banerjee (1964) noted that the Kaimur Group as a whole is more or less uniform in thickness (60-90 m.), excepting a local thickening (~425 m.) at Robertsganj (24° 40' N. & 83° 03' E.). Another maximum of thickness occurs around Chopan. On the basis of this observations, Banerjee and Sengupta postulated two sub-basins in the Kaimur times - a more actively subsiding euxinix eastern basin witnessing deposition of pyritiferous (Bijaigarh shale) and a less actively subsiding open-circulation, normal shallow marine western basin with deposition of quartzarenite and shale.

Summarizing the works of Banerjee (1964, 1974), it may be stated that in the eastern part, the depocenter was near the southern boundary during the Semri time and gradually shifted northward with time. Banerjee (1974) postulated the Vindhyan basin as an inland sea bounded on the north by the Bundelkhand massif and in the south by the Son-Narbada Fault. Ahmad's (1958) isopach form lines also suggest the basin as intracratonic embayment opening towards west. The thickness of the Vindhyans in the Rohtas Plateau has the following geomorphic significance :

(a) The Kaimur sandstone which is 150 metres thick and overlies the Semri limestone works as a strong capping against denudational processes. The variety of landforms over the Plateau top is, therefore, limited. The hard red sandstones and quartzites are very resistant to erosion. Therefore, whatever fascinating landforms are recorded in this thesis, they are mostly due to tectonic disturbances and structural characteristics.

(b) The limestone beds below the sandstones are liable to karstification. Water percolates through joints and cracks, but thick stratum of sandstones does not appreciate a good amount of percolation. That is why the caverns and the associated features below can be peeped through the lower parts of escarpments have taken long time in their development.

(c) It is the great thickness of the sandstone strata that have preserved the numerous caverns even after the great Alpine earth-movements. Had these been a thinner capping, the whole Plateau mass would have been characterized with numerous collapsed karst valley, most of them would have dry.

(d) There is a vast area for the development of Karst landforms. The great thickness of the limestone strata has ample scope for further developments.

2.5 IGNEOUS ACTIVITY

The igneous activity in the Rohtas Plateau has minor geomorphic significance because it has occured after the formation of Basal Stage, the lowest one of the Semri Series. The activities have been recorded by the geologists as Porcellanite stage which is found only on the river bed of the Son in the form of vertical blades. Such features of geomorphic significance canc be observed on the river-bed near Kadhwan and Khairwa situated at the right bank.

The upper beds of the Semri and Kaimur series do not have any imprint of igneous activities. But the activity has some geological impacts on the lower beds.

'The dolerites of Semri series appear to be different from both the Deccan and Rajmahal traps, and may indicate that at the time of the intrusions the total thickness of the overlying beds might have been of the order of 1000' (Ahmad, 1962, p. 537).

2.6 FOLDS

Of the structural characteristics in the Rohtas Plateau area folding is conspicuous in the lower Vindhyan rocks. The forthcoming chapter (Ch. 3.1.2) makes it clear that the beds of Semri series were deposited and were later wrinkled into folds. The Upper Vindhyans which include mainly the Kaimur sandstone overlies unconformably and are tilted. Thus the folded strata do not occur on the surface, but they are clearly observable on the scarp face particularly in the limestone quarries at Tiura-Pipardih, Chunhatta, Baulia, Banjari etc. because the strata have been exposed due to faulting and uplift.

The earlier structural interpretation of these Lower Vindhyan rocks has been modified considerably during the present study. The earlier workers reported a single phase of deformation, resulting in the development of a number of ENE - WSW trending anticlines and synclines. This interpretation has been modified because it is not a single phase of deformation but there is the repetition of the bands and the structural closer formed by these bands. The south facing scarps are showing such repetition in the Lower Vindhyan rocks.

The folding in the Lower Vindhyan rocks has geomorphic significance in the alignment of the limestone caves in the Plateau. There are numerous caverns which are open towards the Son river and the northern plains. A close observation of numerous caverns reveals that the caverns are not at the same horizontal plane but they are following the arcuate oaths and synclinal and anticlinal portions within the Plateau masses. For example, the Chunhatta caverns follow the synclinal paths whereas the Gupteshwar caverns follow the anticlinical portions. Thus the Lower Vindhyan rocks have multifolding system.

2.7 FAULTS

The major structural element of the Rohtas Plateau is the Great Boundary Fault. There are a number of their subsidiary faults, parallel as well as oblique to the major fault but genetically connected with it. These faults can be observed in the southern, south-eastern and northern margin of the Rohtas Plateau.

Although the earlier workers have not any significant mention about the Kaimur and underlying Rohtas Stage, a detailed study by various agencies including Directorate of Geology, Govt. of BIhar has made the following observations regarding the structural involvement of the rock types under study. Shri T.N. Mishra, formerly a Senior Geologist in Bihar Govt. and at present Asstt. Director, Bihar Mineral Development Corporation, Ranchi has made a detailed study of this area. The discussion which follows is based on the talk with him.

i) A major boundary fault running almost in east-west direction from Jaradag Village in the west to Akbarpur in the east has been responsible for the formation of the Kaimur escarpments.

ii) There has been a number of cross faulting which are generally marked by the small *nalas* through which water from the Plateau area is drained into the river Son. The cross-faulting has been responsible for the gradual downthrow of the upper rock formation from west-east. This is why the scarp sandstones occuring an altitude of 2200' in Jaradag area have come down to the ground level near Sasaram area. This is also confirmed by the position of the Upper limestone belonging to Rohtas Stage horizon, which occurs at an altitude of 1200' near Pipardih and same band of limestone comes down to the ground level in Rampur quarry near Ramdihra.

The cross faulting is very evident along Tutla fall near Ramdihra (24^0 46' N. & 84^0 02' E.) where throw of the fault has been measured to the tune of about 50 metres. This has been confirmed by the officers of the Directorate of Geology, Govt. of Bihar, in course of drilling for limestone in and around Rampur quarry. Similar effects have been observed in Ausane Valley near Akbarpur, Bansi *khoh* near Madhkupia, Dumarkhoh near the Mirzapur boarder, Ramkhoh near Jadunathpur.

'Walking along the horizons, one would come across a significant nala channel along the scarp face and the limestone bed-the entire 60-70 m. thickness of it would suddenly disappear. Half a mile or so further ahead it would reappear with equal suddeness' (Ahmad, 1962, p. 517).

There are a number of fault aligned parallel to the main scarp and with downthrow generally away from it.

'It is often difficult to determine when a particular fault was initiated, but later movements along it, if the area comes to form part of a basin area usually easily known. In the present case there was certainly no activity while the Vindhyan sediments were being deposited. This is indicated by the lack of any unconformities within the Vindhyan formations, as pointed out earlier. But it appears, difficult to say when, since the early Cambrian times, this intense zone of faulting came into being. Evidence discussed above suggests that all this faulting, weathering and scarp formation continued into the Recent times, and were connected phenomena. It is possible that faulting was initiated in the Pleistocene times in the Narmada Valley and extended eastward. The sub-recent times the Son has cut its channel deep and was forming the terraces, it has since been rejuvenated' (Ahmad, 1962, p. 521).

The faulting has resulted in the origin of number of interesting features of geomorphological interest. The most impressive tectonic feature of the area in the Kaimur Scarp (Plate : 2) from the top of which one has to descend along a sheer face for up to 400-500 metres. This scarp forcefully impresses any experienced geologist that faulting must have played a significant role in its genesis.

The cross-faulting has originated some deep steep sided river-valleys which intrude into the Plateau mass towards their source-heads and join the Son in the south and south-east. In the north too, such valleys originated by cross-faulting can be observed.

2.8 JOINTS

It is true of rocks in general that joints open the way to weathering (Dury, 1976, p.9). Joints, in other words are actually bedrock cracks along which water can move easily to promote rock decay (Aghassy, 1970, pp. 286-298)

The Rohtas Plateau has numerous joints which are assumed to be mainly produced during the Tertiary earth-movement. 'The Vindhyans are very strongly jointed' (Auden, 1933, p.195). A discussion of joint directions has been made with the hope that some idea would be formed of the nature of the stress which have affected these rocks.

For the purpose, the author has prepared a map of joints following the method of Prasad (1978-79). With the help of set-squares, master set of joints has been drawn considering the master streams, almost parallel to each other. Secondary and Tertiary joints have been drawn on the basis of tributaries. The author has gone to the field to check the jointing patterns. He found them correct in most cases.

In the Plateau of Rohtas, the joints have been drawn as Primary, Secondary and Tertiary joints. The main alignment is from southwest to northeast. Some of the Primary and Secondary joints align from northwest to southeast. These joints are well traversed by numerous tertiary joints. We know that the Tertiary earth-movement has uplifted the Plateau with a tilting towards north-west. Moreover, the northern portion has worked as a **'hinge-zone'** . Consequently the Kaimur sandstone beds are dipping towardsa north and northwest. This is evident from the present surface elevation (Fig. 5.1) which accounts 506 to 600 metres on the southern margin, 212 metres on the north-eastern margin and 225 metres on the north-western margin.

The primary and most of the secondary joints are, therefore, aligned at the right angle of the dip of the bedding planes. Such alignments have been caused by the stress or compression from the south-east and south-west direction during Tertiary earth movements. The recent concept of Plate Tectonics reveals that there were tensional forces from the Arabian oceanic ridges and 90^0 E. ridge of the Bay of Bengal which gave stress to the Peninsular India from both sides to move northwards. Such stresses have resulted in jointings not only in the Rohtas Plateau, but in the whole of the Peninsular Plateau.

The tertiary joints have haphazard alignment in so much so that they are transversed to the primary and secondary joints. As a result these have been a dendritic pattern of joints at many places.

The joints of the Plateau have clear-cut impact on the development of the drainage system of the area and there by the evolution of the Plateau, chapter 4.3 presents such details.

The study of joints and surface drainage shows some controls of joints over the drainage system of the Rohtas Plateau. Strikes and the relative density of the master set of joints seem to control the direction of some streams, their patterns and density in addition to the major control of underlying cavern system that has caused the origin of steep sided constricted and elongated valleys due to subsidence or collapse. The major joints run south-west to north-east and south-east to north-west. In all cases the arrangement of joints fits with the consequent and subsequent streams. The major rivers of the Rohtas Plateau follow the primary or major set of joints particularly in the upper reaches. Tributary streams seem to follow the secondary and minor joint-systems.

The following relationship can be made between the jointing stream and drainage texture

(1) Joints are numerous, both large and small in the Karamnasha, Sura I, Sura II, Karsotha, Koel, Goptha basins where drainage texture is moderately fine and fine.

(2) Though the drainage texture is moderately fine in the Dhoba basin but joints are less. This is due to surface of high gradient on the scarps and notches.

(3) In the south western pattern portion below the scarps the drainage texture is fine but the joints are less. Here the high gradient has been responsible for the development of small

ephemeral streams which are mostly parallel, sometimes without going into the main streams or to have a system.

(4) Joints are less in the north-eastern part of the Plateau where drainage texture is cypher.

(5) The jointings have apparent control on the development of small streams on the Plateau-top, but the streams on the scarps, notches and elongated and constricted valleys are not guided by the jointing pattern, they are controlled either by the gradient or the collapse of the caverns.

2.9 SOIL STRATIGRAPHY

A discussion on the soils of the Rohtas Plateau in respect of its geomorphic evolution has been necessitated for stratigraphic purposes. We know that stratigraphy is that branch of geology which deals with the sequence, age and correlation of rock bodies. 'Soil stratigraphy involves the use of soils in defining a local stratigraphic succession, estimating the ages of units associated with the soils, and suggesting short - or long-range correlations' (Birkeland, 1984, p.326). Since 'soil is a natural body, differentiated into horizons of minerals and organic constituents, usually unconsolidated, of variable depth, which differs from the parent material below in morphology, physical properties and constitution, chemical properties and composition and biological characteristics' (Joffe, 1949, p.662), a knowledge of processes responsible for profile development helps in reconstructing the landscape and delineating the factors that prevailed when the soil in question formed.

The study area has not yet been studied in detail by the Soil Survey Department. The author could collect only some of the drill logs along the axis of Karamchat Dam which is located on the margins of the Plateau, just below the northern escarpment. Besides, a description of the horizontal distribution of soils over the Plateau is available through the articles and reports. The author has also intimate knowledge of the area. All these reveal that there are two distinct sectors in the study area.

2.9.1 DOMINANCE OF RED SANDY SOIL ON THE PLATEAU TOP

'The soil varies from place to place. Sometimes it is coarse and gravelly, at others it is a stiff and shallow red clay, highly, ferrugenous at times passing into laterite and pisolitice iron-ore' (Ahmad, 1965, p. 75). This soil overlies the well-defined Rohtas Plateau and is mostly composed of residual soils which have a rich deposit of vegetable mould. Soils are usually sandy, except in the saucer-shaped valley which have a rich deposits of vegetable where the soils are mature except on marked slopes and bottom of hollows. Ferrugenous content is often considerable. As a whole it forms a thin layer overlying the rocky sub-stratum.

2.9.2 DOMINANCE OF ALLUVIUM IN THE RIVER-VALLEYS

The river valley and other depressions, small and large, are the areas where alluvium predominates. Limestones and sandstones being the predominant parent rocks of the region the soils formed have good proportion of calcium carbonate, silica and some soluble salts. Thus a profile study of the soil worked out and deposited by the river Durgawati reveals that the soils are of heavy to medium texture throughout the profile and contain high amount of soluble salts, iron and calcium carbonate. Soil deposits over the river Karamnasha have sandy texture on the surface followed by loam or clay loam soils below. Some concentration of calcium carbonate are found in the lower depth; soluble salt content is low throughout the profile.

The study reveals that the Rohtas Plateau consists of soils of **in-situ** origin for the most parts on the Plateau top. The hard red Vindhyan sandstones are resistant to weathering. Except in the depression and **nalas**, drifted materials are rarely found. This indicates that the area has been under humid tropical climate in the recent geological past.

As regards the presence of sandstone and sandy clay and calcareous materials in the lowlying river valleys, particularly in the peripheral areas below the escarpments, the author is of the opinion that it is a case of collapse of former caverns. The sandstone angulars boulder of large size are present on the river-beds and banks. These materials cannot be drifted, but they are collapsed materials. It suggests that the Plateau had numerous caverns in the geological past also before the Tertiary earth-movements. They have been collapsed during the Himalayan orogeny developing the narrow deep constricted river valleys (Fig. 5.1).

3

GEOMORPHIC PROCESS

Like many geomorphical studies the central theme of this discussion is to understand temporal and spatial variations in geomorphic processes and the resulting pattern of forms in the Rohtas Plateau. It is emphasised because landforms embody the effects of both contemporary and past geomorphic processes. Geomorphic processes or events in the study area have been spatially contiguous, but have varied with regard to the nature, intensity, frequency and magnitude of occurence over different lengths of time. Consequently, the resulting forms are equally variable and complex in nature.

The processes that create and modify landforms anywhere on the earth's surface in general and the Rohtas Plateau in particular are basically of two types :

1. The endogenic or tectonic processes emanating from the earth's interior; and
2. The exogenic or external gradational processes, which reflect in the main climatic and related influences.

The first set of processes in the Rohtas Plateau are very important in their own right and, infact, influence the second type. Virtually landform cannot be properly interpreted unless the geomorphic history of its own evolution is understood, although it is a fact that the more distant the past the more difficult it is to explain the landforms in terms of evolutionary processes. Many forms of the Rohtas Plateau, However, bear the imprint of processes of the previous geological period. Therefore, the author has tried his best to elucidate the geomorphic history of the region in terms of geologic and evolutionary processes.

The exogenic or external gradational processes effect the breakdown of the basic geology of rocks; erosion and transportation of the weathered or loose surface material; and finally the deposition of the transported material. All these processes operate at or near the earth's surface to produce the landforms which geomorphologists study. The author has, therefore, investigated the interrelationship between the present-day processes and landforms by the field and laboratory techniques. The study area falls under humid tropical climate. Therefore, the exogenic processes include mainly weathering and mass-wasting erosion and transportation by running water and karstification by underground water. The important conditioning factors are climate and vegetation. The climatic parameters like temperature, rainfall, relative humidity and evatransportation play very important roles in determining the nature of the exogenic processes and the character of the resultant landforms in the Rohtas Plateau. Similarly vegetation shields ground surfaces from rainsplash erosion and surface wash; anchors retains regolith on slopes through the action of roots, improves soil structure through the supply of organic matter; and contributes organic acids to the chemical weathering process. A detailed discussion of these conditioning factors have, therefore, been made before the analysis of exogenic processes.

3.1 ENDOGENIC OR EVOLUTIONARY PROCESSES

Geologically and stratigraphically the Rohtas Plateau is the easternmost continuation of the Vindhyan system which extends from Rajasthan in the west to Bihar in the east. The study area of the Rohtas Plateau is interesting in the sense that it becomes a type area of the Vindhyan formations because it consists of very conspicuously and exceptionally both of the Lower and Upper Vindhyans at the same site. It has, therefore, been easier to trace out the evolutionary stages of various formations through the ages, and also to see the geomorphic sequence of the area. It is noted that from the Bank of the son river one can clearly observe beds of limestones, shales, conglomerates, sandstones etc. on the south and south-east facing steep scarps. One can also observe the folding and disturbed or interrupted strata of shale and limestone in the lower portion and more or less horizontal and tilted strata of the upper one. At some places igneous intrusions can also be observed. The geological study of all these strata and structural characteristics has facilitated in tracing out evolution or historical development of the Rohtas Plateau.

The whole formation is said to be the Vindhyan system which is one of the oldest sedimentary formations on the earth. The Vindhyans are the youngest which formed during the late Archaean or early Palaeozoic era or the Purana era of Indian stratigraphy after the Dharwar and Cuddapah sedimentary formations. From the point of view of lithology and structure of the constituent beds the Vindhyan system of Rohtas Plateau has been divided into a (i) Lower and an (ii) Upper subdivisions. The Lower Vindhyans rocks are mainly calcareous and argillaceous in composition and have been subjected to folding. The Upper-Vindhyans are, however, made up of undisturbed and more or less horizontal layers of arenaceous sediments. The two subdivisions are separated from each other by a distinct unconfirmity. All these geological records help in tracing out the geomorphic evolution in the following manner.

Since the Vindhyan system lies unconformably above the older formations and the Cuddapah are absent from the area, it can be inferred that there might have been a long narrow synclinical portion over the original basement rocks in the Central India during the Pre-Cambrian times. 'Post-Bijawar granites-granophyres mark the terminal phase of this cycle (Bijawar). This was followed by development of erosion surfaces, which formed the Pre-Vindhyan basement' (Prakash & Dalela, 1982, p. 58). Thus, 'Vindhyan beds lie upon rocks believed to be ancient gneisses' (Ahmad, 1962, p. 517). Over such basement sedimentation took place.

Sedimentologic analysis of the basin characteristics (Banerjee, 1964, pp. 189-204), depicts the evolution of the basin as a stable platformal shelf bounded by basement ridges. Thus the fault at the boundary must be accepted as having developed later. Another geophysical and subsurface study through boreholes described the Vindhyan basin as a syneclise and not as a graben (Sastri et. al, 1971, p. 224). The strata of calcareous and argillaceous sediments in the lower part of the system speak of the deposits under a marine environment. The deposition took place up to a thickness of 900 metres for a long time until probably the Charnian earth-movement caused the orogeny of strata resulting in fold mountains of repetitive sequence of sandstone, shale and limestone (Mathur, 1981, p. 2). These strata are stratigraphically known as the Semri Series or the Lower Vindhyans. The Semri Series has been traversed by dykes of basic composition, 'The Semri are intruded by basic lavas and show much disturbance and folding' (Krishnan, 1953, p. 18). The Lower Vindhyan range was subject to denudation and in the later periods to subsidence or transgression of sea to the extent that the portion remained as shallow water body and created lacustrine or fluviatile conditions during the Upper Vindhyan times. This period witnessed deposits of arenaceous sediments up to a thickness of 1050 metres lying unconformably over the Semri group. This group of sediments has been divided three series - Kaimur, Rewa and Bhander in the succession from bottom to top. Opinion differs whether all these three series existed in the Rohtas Plateau in the geological past. Presently only the kaimur series is existing and there is no sign of Rewa and Bhander series over the Plateau. At best one can say, as proposed by Ahmad (1962, p. 513), that Rewa and Bhander groups of rock have been denuded. Whatever the case may be, it is assumed with geological and topographic evidences that there was a gradual uplift of the epeirogenic type of the sedimentary strata without any major disturbance in their horizontility. The uplift created, therefore, fault escarpments which are still existing. Thus the scarps of the Rohtas Plateau are as old as the Vindhyans.

Most of the geological records, e.g. Oldham (1859), Medlicott (1859), Medlicott (1859), Mallet (1869), Oldham, Vredenberg & Datta (1901), Vredenberg (1906), Jones (1909), Auden (1933), Dunn (1941), Muktinath (1948), Mishra, (1948, 1969), Ahmad and Mathur (1952), Ahmad (1962, 1962, 1971), Krishnan (1953, 1982), Datta (1953), Pascoe (1959), Krishnan & Swaminath (1959), Wadia (1961), Sahni (1962), Lahiri (1964), Ray, Sah, Seth and Goswami (1966), Mathur, (1979, 1976, 1981), Mukerjee (1973), Srivastava & Iqbaluddin (1976), Ray (1978), Singh (1980), Chanda & Bhattacharya (1982), Prakash & Dalela (1982), Valdiya (1982) and Ray & Bhattacharya (1982) are silent over the happenings occured on the Vindhyan formations of the Rohtas Plateau during the Palaeozoic and Mesozoic eras which witnessed world-wide earth-movements like Caledonian, Armorican and Hercynian in the Sulurian, Carboniferous and Permian periods respectively. The Palaeozoic and Mesozoic eras together cover about 800 millions of years. It is somewhat unconvincing that the Plateau area could not witness any happenings. The statements appear, therefore, to be unacceptable when one says, 'the area did not undergo any uplift since Cambrian or Vindhyan times up to the Tertiary era'. (Ahmad, 1965, p. 24). The author endeavours, therefore, to present a hypothetical evolutionary picture of the Rohtas Plateau since Cambrian. It has already been noted that the two major groups of the Vindhyan consisting of the Semri, Kaimur, Rewa and Bhander were found during the Pre-Cambrian periods having a conspicuous unconformity marking the folded, intruded and disturbed strata in the Semri series and the comparatively horizontal strata of the Upper Vindhyans. The author presents two alternatives of the evolutionary stages.

3.1.1 PROBABLE EVOLUTION NO. I

Let us assume that both the Lower and Upper Vindhyans were raised to a considerable height above the Vindhyan basin through faulting on north and south directions. The Bhander series was subject to denudation, it was removed completely or partly and the Plateau was lowered; but it was uplifted due to Caledonian earth-movement. Again the denudational processes worked on the Plateau and the Bhander series was completely removed and the Rewa was partly denuded. The Hercynian earth-movement raised the Plateau again, it was subject to denudation causing complete removal of Rewa series and the part removal of the Kaimur series. The Tertiary earth-movement raised the Plateau up to the present height.

If we accept this assumption we can face the following objections :-

(i) Had there been an uplift of the whole of the Upper Vindhyans and part of the Lower Vindhyans the time span of even Cambrian to Silurian periods would be sufficient to erode down the Upper Vindhyans and thus there would be no trace of the Upper Vindhyan rocks, rather it would have been a considerable removal of the Semri series. But it is not the case.

(ii) The assumption that both the Lower and Upper Vindhyans rose above the base level is untenable because in that case the time span from Cambrian to the Triassic was sufficient for solution of the calcareous deposits leaving no imprint of the Semri series on the date.

We are, therefore, bound to go for an alternative proposition :

3.1.2 PROBABLE EVOLUTION NO. II

On the basis of the analysis of sedimentary strata of both Lower and Upper Vindhyan formations found in the study area of the Rohtas Plateau, we can assume that the Vindhyan geosyncline or 'syneclise' (Sastri et.al, 1971, p.222) was 'a stable platformal shelf bounded by basement ridges' (Banerjee, 1964' p.189). It was completely a marine condition under which deposition of calcareous and argillaceous materials took place in so much that the formation covered a thickness of 900 metres of compressed material particularly in the Rohtas Plateau area (Fig. 3.1). This formation is known as Semri series in geology (Table: 3.1). It must be a long duration of the Archaean or Pre-Cambrian era before the last Archaean earth-movements i.e. Charnian could fold the strata into mountains chain.

Thus there was the first uphevel of the Vindhyan formations above the base level. It can be assumed that the moment folded strata rose above the sea level denudational work would have been started. The unconformity between the Semri and the Upper Vindhyan rocks reveal that the upthrown folded strata might have been denuded to the base-level.

The arenaceous character of the Kaimur sandstone lying over the Semri helps to infer that there was a shallow water body either in the form of lagoons or fluviatile condition over the denuded Semri Groups. It existed either due to subsidence of the area or transgression of the sea. Whatever the case may be, there existed a sinking water basin of the Semri base over which sedimentation of arenaceous materials took place for a long time (Fig. 3.1). It was alternated with dry or semi dry and more humid climate to facilitate the formations of sandstones and shale of the Kaimur series, shale and sandstone of the Rewa series and shale, limestone and sandstone of the Bhander series (Table : 3.2) All these three series are grouped into the Upper Vindhyans. Though the Rewa and Bhander are absent at present from the Rohtas Plateau, their existence over

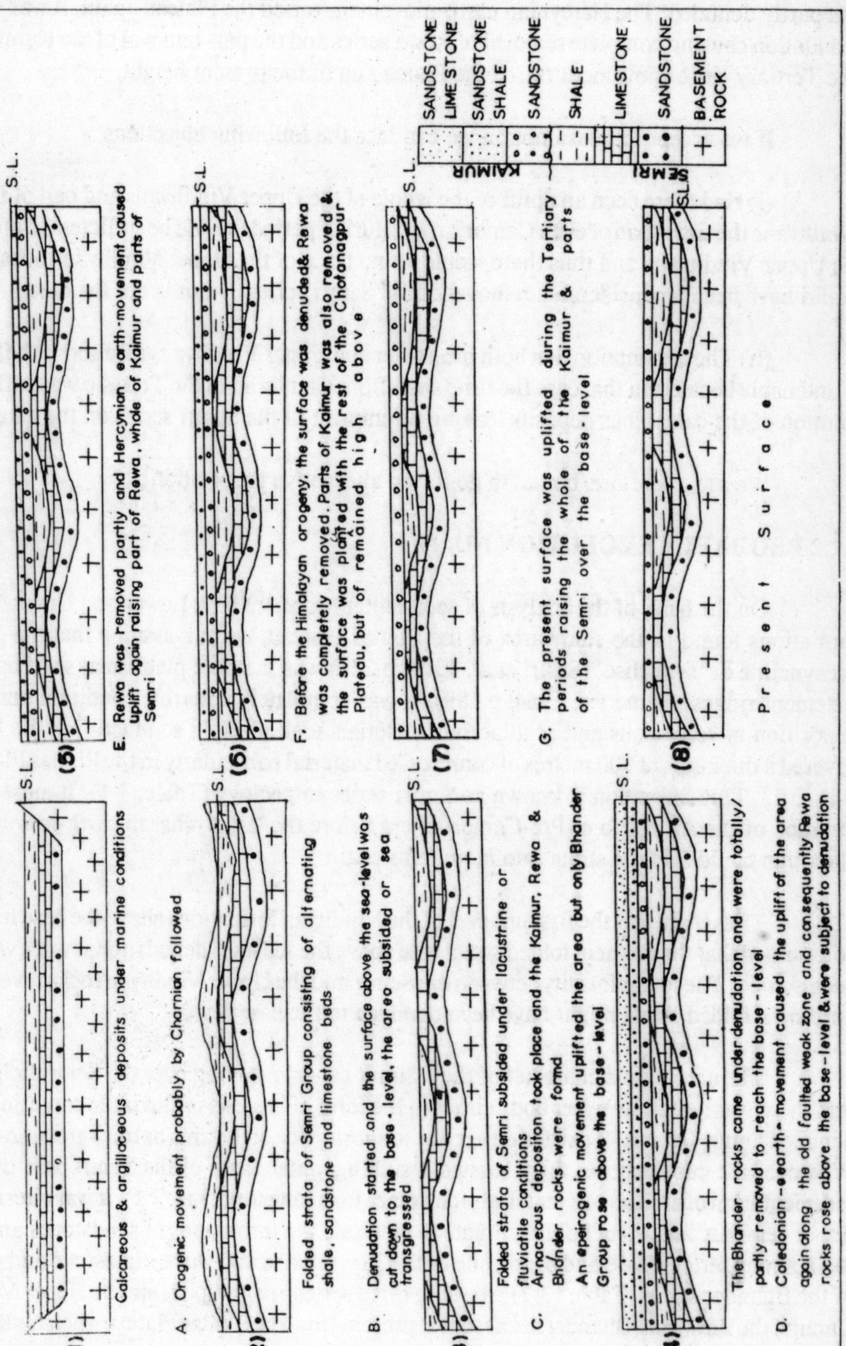
GEOLOGICAL EVOLUTION OF THE ROHTAS PLATEAU
(1)
S.L.
Calcareous & argillaceous deposits under marine conditions
A. Orogenic movement probably by Charnian followed
(2)
S.L.
Folded strata of Semri Group consisting of alternating shale, sandstone and limestone beds
B. Denudation started and the surface above the sea-level was cut down to the base-level, the area subsided or sea transgressed
(3)
S.L.
Folded strata of Semri subsided under lacustrine or fluviatile conditions
C. Arnaceous deposition took place and the Kaimur, Rewa & Bhander rocks were fomd
An epeirogenic movement uplifted the area but only Bhander Group rose above the base-level
(4)
S.L.
TheBhander rocks came under denudation and were totally/ partly removed to reach the base-level
D. Caledonian-earth-movement caused the uplift of the area again along the old faulted weak zone and consequently Rewa rocks rose above the base-level & were subject to denudation
(5)
S.L.
E. Rewa was removed partly and Hercynian earth-movement caused uplift again raising part of Rewa, whole of Kaimur and parts of Semri
(6)
S.L.
F. Before the Himalayan orogeny, the surface was denuded & Rewa was completely removed. Parts of Kaimur was also removed & the surface was planed with the rest of the Chotanagpur Plateau, but of remained high above
(7)
S.L.
G. The present surface was uplifted during the Tertiary periods raising the whole of the Kaimur & parts of the Semri over the base level
(8)
S.L.
Present Surface
S.L.
KAIMUR
SANDSTONE
LIMESTONE
SANDSTONE
SHALE
SANDSTONE
SEMRI
SHALE
LIMESTONE
SANDSTONE
BASEMENT ROCK

FIG. 3.1

the Kaimur has probability otherwise we cannot explain as to what happened in the long geological times from the Cambrian to the early Tertiary.

TABLE : 3.1

THE LOWER VINDHYAN

Limestone	1 A	Cycle V
Shale	1	
Limestone	1 A	Cycle IV
Shale	1	
Sandstone	2	Cycle III
Shale	1	
Limestone	1 A	
Sandstone	2	
Shale	1	
Porcellanic shale		
Trappoid beds	1	Cycle II
Porcellanic shale		
Limestone	1 A	
Sandstone	2	Cycle I
Conglomerate	2 A	

After Mathur, 1931, p.2.

Thus, there were three successive formations of the Upper Vindhyans over the Semri series. The three formations were separated by diamondiferous conglomerate beds (Mukerjee, 1973, p.330). Now let us assume that the first epeirogenic movement occurred before the Cambrian which caused great boundary faults along the northern and southern margins of the basins and raised the Bhander beds above the base-level. This uplifted portion was subject to denudation and was totally/partly removed. The Caledonian earth-movements of the silurian - Devonian period could not wrinkle the strata owing to their hardness but they could have raised the portion to a considerable height along the old faulted weak zones and consequently the Rewa series with some portions of the Bhander might have come above the base level of the old geosynclinal basin.

TABLE : 3.2

THE UPPER VINDHYAN

	Upper Bhander Sandstone	2
	Sirbu Shale	1
Bhandar	Lower Bhander Sandstone	2
	Bhander Limestone	1 A
	Ganurgarh Shale	1
Rewa	Upper Rewa Sandstone	2
	Jhiri Shale	1
	Lower Rewa Sandstone	2
	Panna Shale	1
Kaimur	Upper Kaimur Sandstone and conglomerate	2
	Bijaigarh Shale	1
	Lower Kaimur Sandstone	2

After Mathur, 1981, p. 2.

The period between the Caledonian and the Hercynian earth-movements was sufficient to reduce the Bhander and Rewa formations and the surface might have been lowered again. The Hercynian earth-movements of the long and vast faulted basins over the Indian Peninsula might have affected this weak zone and the Kaimur along with portions of the Rewa above and the Semri below would have uplifted through the faulted margins. The period between the Permian and the Eocene can be regarded as a silent age in regard to any epeirogenic or orogenic movements in the study area. There might, have been denudational processes which could remove the Rewa completely and the Kaimur partly. This period also witnessed karstification in the Semri series which was overlain by the Kaimur sandstone. It can be assumed that numerous karst landforms including large caverns might have developed. The time was sufficient for it. It can however be noted that the study area was not reduced to a peneplain as could be seen the adjoining Cotanagpur portions. It remained high above the surface. It might have planated.

The tertiary earth-movement which was responsible for great Himalayan orogeny has scars epeirogenic movements over the Plateau and the adjoining Chotanagpur Highlands. The Rohtas Plateau was further uplifted along the boundary faults, it gained an ultimate height, the Semri was further thrown up above the basin level and numerous caverns collapsed resulting in separation of Plateau portions from the main mass through karst valleys and development of long narrow steep sided deep valleys known locally as khohs.

Thus the Rohtas Plateau and its fascinating landforms are the result of tectonic movements through the ages. One can, however , emphasize that the existing landforms are largely due to tectonics of the Tertiary era. The fascinating landforms are the numerous caverns and the khohs which appear to be originated due to collapse of caverns. The Tertiary tectonics become important in the sense that no previous movements found existence of caverns owing to the non-existence of the Semri series above the base level. This assumption gets support from Ahmad (1962, p. 517) when he says, 'it, then, seems reasonable to the suspect that the collapse of the caverns has not been due to the weight of the overlying beds, but has been brought about by some tectonics disturbance. This view is supported by the fact that all caverns have collapsed, whereas normally some would probably, have escaped. This envisaged disturbance could only be very recent for the agents of erosion have not had the time to remove the shattered overlying quartzites and to re-expose the cavern openings'.

As regards the present height of the Plateau and its tilting towards north and east there are explanations. We know that the adjoining Chotanagpur rose to a maximum heights of 1150 metres while the maximum height of Rohtas Plateau is only 610 metres. Both the Plateaus were subject to uplift during the Himalayan orogeny. But the uplift of the Rohtas Plateau in relation to Chotanagpur highland is very low. It is because of the fact that the force is applied from two sides there must be a point where both forces are neutaralised. The force which uplifted the Himalayas came from the two sides that was Peninsular India on one hand and the Chinese plateau on the other hand. Similar is the case with Rohtas Plateau. The Rohtas Plateau was left near the hinge zone or where the disturbance occured but to a lesser extent. 'It is interesting to note that along latitude 25^0 there was practically no movement, this being of the nature of a hinge zone located somewhere in the middle of the southern limb of the trough' (Krishnan, 1953, pp. 58-59), north of the Rohtas Plateau. Naturally the area near hinge zone was less uplifted. Further the tilting towards the hinge zone is natural.

3.2 EXOGENIC OR EXTERNAL GRADATIONAL PROCESSES

3.2.1 CONDITIONING FACTORS

3.2.1.1 Climate

The discussion of the climatic features of the area under study has been necessitated owing to the fact that any change, creation, destruction and development of sequential landforms is accomplished by a moving medium, **i.e.,** flowing water, the movement of ice or the wind. And all these important exogenous processes are closely related to the climatic conditions. We know that the moment landforms are created as initial landforms by endogenous processes; destruction and erosion proceed, tending towards a final levelling. A preparatory stage to this erosion consists of the weathering and loosening of the rock, by mechanical or chemical means under certain climatic conditions. Moreover, erosion of ten continues basically under the influence of gravity, which moves masses downwards in mass movement or displacements of the ground. It affects both loose materials and rock masses in situ. Through these mass movements on more or less tilted plains, parts of the earth's surface can be laid bare by moving media and fresh rock weathering can be initiated (denudation). And through these dynamic processes, the sequential landforms are developed under certain climatic conditions. Thus, the role of climate in the development of landforms has been greatly emphasized by numerous geomorphologists, particularly by Penck (1905, pp. 165-174), Chorley (1957, pp. 628-638), Pecsi & Szilard (1968, pp. 17-28), Cotton (1969, p. 354), Martonne (1973, pp. 61-75), Stoddart (1969, p. 163), Tricart & Cailleux (1972, p. 295), Budel (1973a, pp. 104-130), Louis (1973, pp. 153-170), Passarze (1973, pp. 91-95), Derbyshire (1976, p. 1), Prasad (1980, pp. 385-405), and many others. They hold the opinion that 'the ability of the rocks to witnessed erosion does not depend on their physical but rather on their **morphological hardness**. This varies according to climate. Granite, for example, is one of the hardest rocks in mid-latitudes whereas in the moist tropics, where deep weathering prevails, it is soft and sand like' (Budel, 1973b, p. 203). Similarly epeirovariance is not, as a whole, a determining factor of the relief features. Epeirovariance governs the ultimate height and general steepness of slopes. Of course, 'the form of an uplifted block or a step produced epeirogenetically appears in its pure form only in the relief of the ocean bed. On land, the effect of exogenic erosion processes begins at once, changing the 'raw block' into a greatly reduced, many-membered relief of flat surface, ridges, slopes and valley which we see today' (Budel, 1973b, p. 204). Thus the various meteorological phenomena, e.g., temperature, precipitation, frost, snow-cover, wind, etc. are the main climatic elements and all these are significant exogenous factors for weathering and denudation and thereby for the development of landforms. At higher temperature, chemical, biochemical and biological weathering become increasingly significant, rating doubling over every 10^{0}c. rise in temperature. In areas where period of high temperature is followed by general rain, the process of weathering and its removal are both equally important. Thus each climatic element has significant role in shaping the earth-surface. The Rohtas Plateau has experienced both temporal and spatial variations in climatic conditions. The present day Tertiary landforms are, no doubt, the product of climatic phenomena. Before we make a relationship between climate and landforms of the Rohtas Plateau, let us examine some of the characteristics of climate of the area under study.

Climatic Features :

Since most of the parts of the area under study is inaccessible even today (Ch. 1.6.2) no meteorological station could be set over the Plateau itself. Therefore, whatever data have been used for such analysis, they are collected from the sations located either in the peripherial areas of the Plateau or to the nearest points like Indrapuri, Dehri etc. The author has, however, visited the area many times and has experienced the climatic conditions of almost all the seasons. Thus the data have been verified by personal experiences and observations.

The climate of the Rohtas Plateau is generally dry and bracing for the greater part of the year. The cold season starts early in November and extends to late March. The hot season is from April to about the middle of June when the monsoon season commences. The monsoon continues till the end of September. October is a transition month.

The Rohtas Plateau lies in the interior far from the sea. Therefore, it is natural to get greater extremes of climate than the regions near the sea. From the data (Table : 3.3) it can be said that in the hottest month of May the mean maximum temperature during the day is 41.6^{0}C and the mean daily minimum is 27.7^{0}C. In January which is the coldest month the mean minimum temperature is 23.9^{0}C and the mean daily minimum is 11.4^{0}C. In summer months temperature

TABLE : 3.3

TEMPERATURE DATA (INDRAPURI BARRAGE) IN C^{0}

Years	1980		1981		1982		1983		1984		1985	
Months	Max.	Min	Max.	Min.	Max.	Min.	Max.	Min.	Max.	Min.	Max.	Min.
January	28.07	4.83	26.13	7.61	26.69	5.17	27.52	1.77	26.18	2.05	28.02	1.77
February	32.24	6.28	33.52	9.28	29.74	5.78	29.52	2.78	28.02	2.27	31.69	2..21
March	43.07	10.50	36.14	11.28	36.02	10.00	38.03	8.06	39.19	6.28	40.03	7.28
April	47.64	17.01	39.53	19.29	39.32	16.79	40.31	13.01	42.81	13.78	42.58	11.58
May	44.53	21.29	44.09	22.62	43.53	18.34	42.53	17.29	45.53	17.29	45.03	14.78
June	41.53	21.29	44.86	21.29	44.53	21.79	45.03	18.01	43.09	17.79	44.53	15.79
July	36.41	21.29	36.14	22.29	43.53	21.01	43.09	17.29	36.52	18.01	36.52	15.79
August	35.08	20.79	35.08	22.51	36.02	21.79	36.02	19.79	35.02	16.06	35.69	14.78
September	34.63	20.79	32.34	21.29	36.02	16.17	35.02	18.34	37.69	14.78	34.36	14.28
October	34.08	18.34	33.52	14.28	35.30	14.28	34.52	9.61	34.69	11.78	33.52	7.78
November	32.08	22.44	31.13	8.50	32.02	6.28	30.52	6.17	33.02	17.80	30.32	1.77
December	30.07	8.56	27.07	5.78	27.68	4.28	27.52	1.77	29.69	12.05	30.19	0.77
Annual	36.61	16.11	31.22	17.89	32.84	13.47	35.56	11.01	35.95	12.41	36.05	9.04

Source : Indrapuri Barrage Office

is as high as 47°C on individual days. While the lowest temperature recorded in the cold months is 4° to 5°C. The Rohtas Plateau region because of its elevation is less hot in summer, mights are cooler and similarly in winter the average temperature is less than the adjoining lowlying plain areas.

Light south-westerly or westerly wind prevails in the winter and early summer months. Towards the end of summer easterlies set in and continue throughout the monsoon season. During the winter and summer months skies are generally clear or lightly clouded but towards late summer cloudiness is more in the afternoon. In the monsoon months skies are heavily clouded or overcast.

Locally heavy rain and strong winds are experienced in the Plateau in association with some of the storms or depressions which move inland from the Bay of Bengal during the monsoon and post monsoon months. Thunderstorms accompanied with squalls occur in the summer months. Even during the monsoon months rain in sometimes associated with thunder. Morning fogs are possible during the winter months.

All the climatic features discussed earlier reveal one important aspect that the area under study is under sub-tropical climate where dry and hot seasons are followed by a wet & rainyseason.

TABLE : 3.4

Annual Rainfall of Rohtas District (Blockwise)

Name of Blocks	Rainfall in mm.								
	1975	1976	1977	1978	1979	1980	1981	1982	1983
Sasaram	748.0	1144.7	1194.7	1742.8	712.1	957.7	700.9	580.9	782.5
Sheosagar	773.0	1296.9	1236.2	1543.9	669.2	791.7	690.7	676.9	909.6
Chenari	628.4	1316.5	1330.5	1722.1	811.3	732.4	874.6	743.9	955.2
Rohtas	890.5	988.8	978.7	1816.4	937.5	1744.5	1023.7	994.5	969.7
Nauhatta	889.0	762.2	1614.8	1675.7	884.8	652.6	961.0	821.0	890.5
Bhagwanpur	756.6	1184.6	929.8	1252.5	681.5	879.7	1145.4	692.9	667.7
Chainpur	622.0	769.6	1065.0	1533.1	468.8	779.5	328.8	1198.1	723.0
Adhaura	1230.1	1646.7	1844.0	2042.2	789.0	1513.3	1290.8	702.5	1329.2

Source : District Statistics Office, Sasaram

This seasonal variation of climate has certain impact on the evolution of landscape. In dry periods, mechanical weathering dominates. Intense insolation by day and strong radiation at night cause scales to separate and break off rock surfaces. Owing to the development of cracks on the Kaimur sandstone in large scale, these has been sometimes a sudden collapse of the materials, even in the dry periods. Such a collapse was observed by the author himself in 1971 near Dabua quarry on the southern scarp-face. Over such mechanically weathered portions the heavy downpours in rainy periods become very much effective.

TABLE : 3.5

Monthly Rainfall in Adhaura Block, 1980 - 1984 (In mm.)

Months	1980	1981	1982	1983	1984
January	0.0	60.5	35.2	8.4	31.8
February	0.0	0.0	29.3	11.4	90.4
March	41.0	41.0	150.4	0.0	0.0
April	0.0	0.0	4.6	47.6	8.5
May	55.0	55.0	86.2	20.2	0.0
June	227.7	122.7	189.2	93.2	369.5
July	538.2	603.4	239.5	320.9	416.6
August	457.3	281.6	409.6	176.9	563.3
September	238.3	235.9	151.3	429.4	133.3
October	40.0	0.0	6.8	153.0	25.3
November	0.0	33.6	0.8	0.0	0.0
December	47.0	8.2	5.8	0.0	0.0

Source : B.D.O. Office, Adhaura

3.2.1.2 Natural Vegetation

Broadly speaking vegetation tends to reduce both runoff and erosion. It is 'perhaps the greatest deterrent to soil erosion for its tend to protect the surface from raindrop impact, reduce the amount of water available for runoff by consuming it and by improving infilteration capacity, and (by increasing surface roughness) decrease the velocity of runoff. In addition, plant roots help to keep the soil in place. A grass cover it often the most efficient defence against soil erosion by running warer' (Cooke and Doornkamp, 1978, p. 39).

The Rohtas Plateau has a considerable area (almost 90.7 per cent) under forest cover. It is true that the gradual elimination of the forest cover by and large to meet the demands of neighbouring areas for timber and firewood has caused immense loss to it. The reports reveal that the 'area must at one time have borne magnificent forest' (Haines, 1978, p. 6). Sir J.D. Hooker records specimens of Hardwickia 120 ft. high as existing even in his time. It is unfortunate that such vegetation cover could not be conserved. Consequently the hilly tracts in the area moderately forested but the growth is not luxurient and there is no large timber left. The woods have bamboo in plenty mixed with a fair amount of sida, persar, khair and ebony sal.

The adjoining map (Fig. 3.2) shows the vegetation cover to date. The topographical sheets prepared on the basis of aerial photographs provide the details of vegetation cover. Also the author has visited the most parts of the area. The study shows that the Plateau is more or less covered either with forests or scurbs. The author thought it desirable, therefore, that the

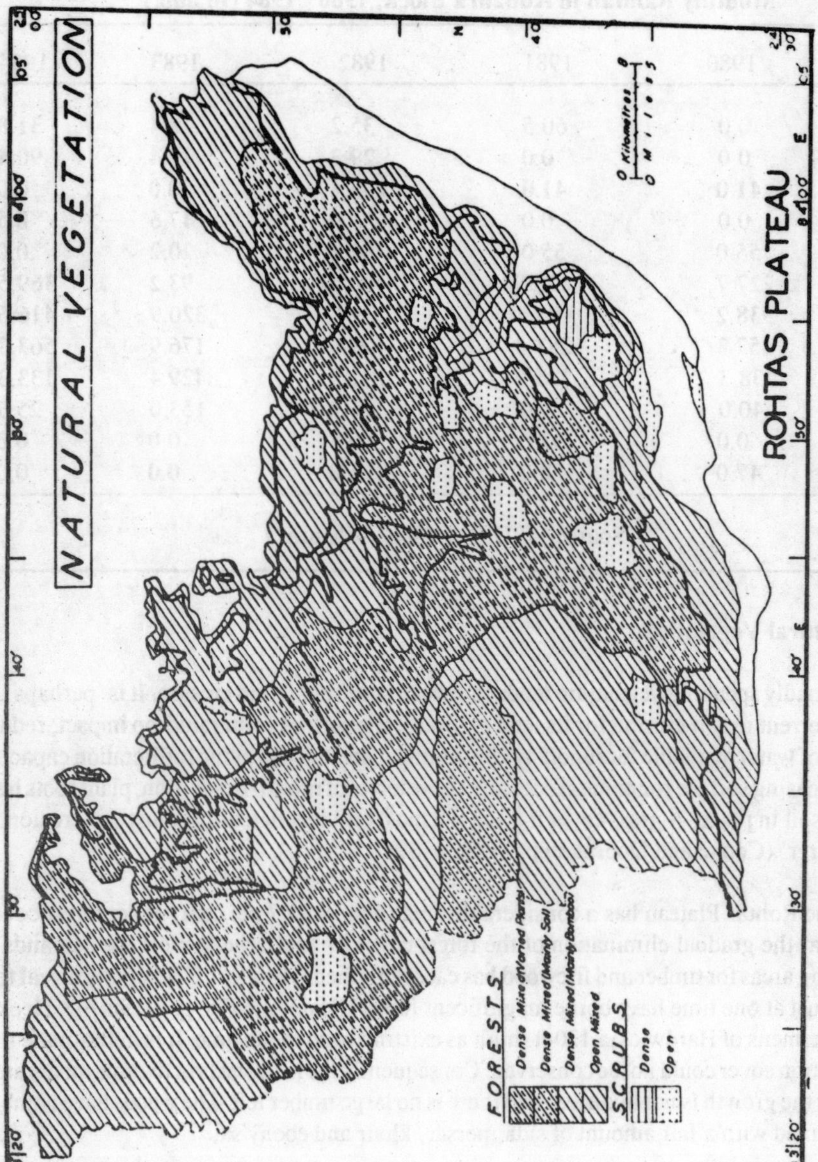
NATURAL VEGETATION
ROHTAS PLATEAU
FORESTS
Open Mixed
SCRUBS
Dense
Open
Kilometres

FIG. 3.2

discussion be made under two major heads (Table : 3.6) :

TABLE : 3.6

NATURAL VEGETATION

Types	Total area	% of Forested area	% of total area
A. FOREST			
1. Dense Mixed (varied plants)	932	43.8	39.3
2. Dense Mixed (mainly sal)	156	7.4	6.6
3. Dense Mixed (mainly bamboo)	168	7.8	7.1
4. Open Mixed	680	31.9	29.0
B. SCRUBS			
1. Dense	16	0.7	0.6
2. Open	178	8.4	7.6
Total	2130	100.0	90.7

3.2.1.2.1 Forest

3.2.1.2.1.1 Dense Mixed Forest (varied plants)

It constitutes one of the most heavily wooded areas of the Plateau. It covers 932 square kilometres area, which is 43.8 per cent of the total forested area and 39.8 per cent of the total study area. Thus forest is represented by various types of plants **i.e.** *khair (acacia catechu), asan (terminalia tomentosa), amaltas (cassia fistula)* etc. This forest covers mostly the eastern and western middle parts of the Plateau top.

3.2.1.2.1.2 Dense Mixed Forest (mainly sal)

This type of forest is represented by mainly *sal* tree with other plants. It covers an area of 156 square kilometres which is the 7.4 per cent of the total forested area. This type of forest is found in the Karamnasha valley and eastern part of the Plateau top.

3.2.1.2.1.3 Dense Mixed Forest (mainly bamboo)

The most important tree is such forest is bamboo. The bamboo mostly covers south-western and eastern parts of the scarp face on debris slope. In eastern portion the bamboos are mostly found in the upper reaches of the Ausane Valley. It covers an area of 168 square kilometres which is 7.1 per cent of the total area.

3.2.1.2.1.4 Open Mixed Forest

The most open mixed forest consists of stunted trees. There are a number of dry rock-loving plants to be seen in this forest. It consists an area of 680 square kilometres. Open mixed forest are found in the north-western, northern, north-eastern, and the outer scarps zone in the eastern portion.

3.2.1.2.1 Scurbs

It consists of trees and scrubs which have, quite often, stunded growth. On the Plateau top scrubs are found mostly in the depression areas. Long grass, chiefly spear-grass kus (poa cymosuoides), khaskhas (andropogen muricatus) with small thorny bushes are found on the Plateau top.

Dense scrubs are found only in the eastern part of Rohtasgarh area while the open scrubs are observed in the depressed areas.

3.2.2. PROCESSES

3.2.2.1 Weathering

The sculpture and evolution of the landforms in the Rohtas Plateau is very much associated with the weathering in the area. We have noted that the Plateau has tilted alternating layers of shale, limestone ans sandstone lying over each other with thick sandstone capping (Ch. 3.1.2). These strata peep only through the steep scarps and steep-sided valleys. The Plateau top is, however, almost homogeneous in structure. We have also mentioned (Ch. 3.2.1.1) that the area experiences seasonal variation as well as diurnal changes in temperature and spatial variation in climatic conditions.

Under such circumstances of materials and climatic features, the Plateau is subject to both mechanical and chemical weathering predominantly of insolation, forest-fire, spalling, biotic, solution, oxidation, carbonation and hydration etc. The running water and gravity are the main agents of removal of materials.

3.2.2.1.1 Physical Weathering

Insolation weathering plays a vital role in the Rohtas Plateau. Temperature changes cause expansion or shrinkage of rocks. A rise in temperature causes a rock to expand, and a fall causes it to contract. Repeated temperature changes cause a rock to break up.

The effect of insolation weathering is more pronounced in the wall-like southern scarps of the Plateau facing the Son river than the top of the Plateau; because the sun's rays are more intense on the scarps. The heat absorbed by the constituent rock minerals in the southern part lead to greater expansion of minerals and joints. Alternate heating and cooling causes rapid disintegration of rocks in the southern part in comparison to the northern scarp face which for the most part of the year remains shadow. The quantity of mechanical weathering processes is, therefore, wide unequal in the southern and northern scarp faces. This is clearly visible in the amount of weathering products which are found on the debris slope area of the southern scarps and the northern scarps. The debris slope in the south is wider and contains more fragmented materials than that in the north. Also elevation of the debris slope is higher in the south.

On the Plateau surface the process of mechanical weathering works in collaboration with chemical weathering. The sandstone which forms the top of the Plateau is exposed at most of the places in which joints are clearly visible. These joints are affected by the daily range of temperature especially in hot summer months. In the succeeding rainy season the joints are filled with water along with the pour spaces of the rock. Since the sandstone consists of iron in large proportions, the wetting process starts oxidation of this minerals, which makes the joints weak and susceptible to easy fragmentation. Thus the Plateau top is weathered by the simple process of peeling or onion like exfoliation. The process of mechanical weathering is not very rapid. The upper portion of the Plateau of hard sandstones responds not very quickly to thermal alteration. Therefore, the upper layer has remained as a protecting layer for the underlying limestone strata for a very long time. Had the upper layer been eroded away by the combined forces of mechanical and chemical weathering, the shape and slope of the Plateau as well as bounding scarps would have changed completely. The lower strata of limestone is in a protected state and as long as the upper sandstones are there, the steepness of the scarps will persists.

The scarp zone is an area where rock-fall is a very important process of unloading. This is somewhat different from landslide and soil creep. In a landslide a section collapses after being throughly rain soaked, which causes the rock particles loose cohesiveness. Similar is the case with soil creep which is generally caused by heavy rainfall. In a rock-fall the weathered material drops down due to gravity sliding without any dependence on rainfall. This may take place in a dry period or in wet period. The products of insolation weathering or fragments produced by fire drop down on the adjoining talus slope.

The effects of forests fire on physical weathering can also be seen on the Plateau-top and on the southern scarp zone. The grasses on the Plateau surface become dry in hot summer. The grasses are fairly long and they easily catch fire in summer. Similarly in the southern scarp zone forests fires break out in the areas where mixed jungle containing mainly bamboo are found. In the month of May and June when temperature is highest and 'Loo' blows from the morning till late in the night the bamboo forests catch fire. The fire is produced by the rubbing of bamboo stems against each other. The author has seen these fires from his native village almost every year in summer months. As the fire spreads the entire scarp zone is gulfed by rising flames and smokes and entense heat is produced. This effects the rocks of the area, very quick expansion of the minerals causes splitting of thin layers of rocks which may be called spalling. All the constituent minerals in a rock do not expand at equal rates. Some expand more and some less. This produces stress in the body of the rock and cracks develop due to unequal expansion. But the fragmentation is definitely caused by this process which is one of the most important causes of physical weathering in the south.

3.2.2.1.2 Chemical Weathering

The chemical weathering of Rohtas limestone has produced extensive passage ways, pits, holes and irregular fluttings on cave walls and ceilings which testify to extensive dissolving power of groundwater. Limestone are made up of calcite ($CaCo_3$) and dolomite (Camg $(Co_3)_2$ which get easily dissolved in rainwater or groundwater containing carbon dioxide. The water containing Co_2 behaves as carbonic acid (H_2Co_3) on the body of limestone and when this water finds a passage through joints and cracks of the rock, the chemical reaction on calcite and dolomite is set in motion which is partly regulated by the variables of temperature and pressure. 'The variation of solubility with pressure obeys Henry's Law, which means that the solubility of Co2 increases with a decrease in temperature and an increase in pressure. Thus, all things being equal, Co, is more soluble at lower than at higher temperatures' (Sweeting, 1972, p. 25).

The sandstone coverings of Rohtas limestone do not permit direct exposure to rainfall but as the rain water percolates downwards through pore spaces and joints the underground water pressure increases and temperature becomes low. Thus an ideal condition for complete solution of limestone is reached and the chemical decay starts. Large number of caves in the body of limestone below the sandstone layers are due to the vigorous nature of solvent action of carbonated underground water. Such caves are numerous along the quarries like Chunhatta, Bariacha, Baulia in southern scarp zone and many river valleys like Durgawati, Karsotha, Sura, Karamnasha along the northern scarps. All these features owe their origin to chemical weathering of limestone which differs completely from the process of rock slip, creep or rockslide in so far as production of limestone caves is concerned.

3.2.2.1.2 Biotic Weathering

The Rohtas Plateau with its outlying spurs is largely covered with forest growth. Slopes are densely clothed with jungle. On the Plateau is a variety of grasses, mainly peargrass kus (poa cymosuroides) and khus khus (andropogan muricatus). The vast stretches of bamboo cover the debris at the base of the cliffs. The forest consists for the most part of a mixed growth of stunted trees of no great height. There are patches of sal (shorea robusta) of which timber is poor. Among other trees that grow amaltas (cassis catechu), palas (botea frondosa), asan (terminalia tomentosa) and mahua (bassia latifolia). Perhaps the most useful of all the trees which clothe the hill and the undulating surface at their base is Mahua.

Natural vegetations act as an important agents in shaping the geomorphic features on the face of the plateau. Vegetation is an imporatnt agent of weathering. It effects the weathering by both mechanical and chemical methods. The growing roots of grasses, shrubs and trees exert an almost incredible force as they work down in crevasses. Roots try to enter downwards in crevasses making wider and wider as roots become longer and longer in size. In this way, cracks are widened by expansion during growth. Wedges of rocks are forcibly shouldered aside. Not only this, when cracks become wider due to the growth of roots crevasses and cracks become the prey of wind and water. These subaerial agents penetrating into greater depth make the rocks to be disintegrated. Very fine example of rock weathering by roots is found in the Rohtas Plateau region everywhere. Due to this process small ditches are found where the rocks are softer but on the slopes the rock is totally shifted aside and the root is seen without rock veneer being weathered by water directly which will lead the tree to fall on the ground ultimately. The larger the tree the greater is the root than smaller ones, so according to their sizes the intensity of weathering varies.

Plants of all kinds including fungi and lichens, also contribute the chemical weathering. They abstract certain elements from rock material and consequently liberate others in the process. By contrast with the exothermic oxidizing nature of most inorganic weathering processes photosynthesis is endothermic and reductive. In addition to carbon, nitrogen, oxygen and sulphur, most plants and bacteria use and immoblize iron, magnesium, potasium and certain other elements. More-over, water containing bacteria attacks the minerals of rocks and soils much more vigorously than it could do in their absence. This leads to the formation of carbonoxide and organic acids together with traces ammonia and nitric acid due to which the solvent power of soil water is increased. Consequently, the water removes the rock mass by dissolving such constituents easily. For this reason the process humus which is the characteristics organic constituent of soil is produced. This humus increases the solubility of water which can dissolve certain substances, such as limestone, which are ordinary insoluble. This is why, in the rainy season, the rivers like the kao, the kudra, the Durgawati, the Karamnasha, the Sura are characterised with rich humus making the colour of water quite red. The protective action of vegetation in preventing the weathering of rocks is also very important influencing the landform characteristics. Rootlets bind the soil into a woven mat, so that it remain porous and is able to absorb water without being washed away. The destructive effects of water are thus effectively controlled. Forests break the force of the rain and control the flow of rivers. All the rivers of the Rohtas Plateau are marked with clear water except during the rainy season. Where the trees have been out by cultivators and graziers, the soil erosion is quite operative. Thus vegetation exerts a major influence on the development of soil characteristics.

Animals influence rocks. Ants, termites, wild dogs, birds and other animals build mounds which locally are very conspicuous. Earth worms, rodents and termites play an important part in preparing material for removal by rain wash.

Man has become the most outstanding boitic agents for weathering. Man-made quarries, road cuts and fills and many other types of excavations profoundly modify the face of the Plateau at local scale. This Plateau has been remodelled by human activities with peculiar geomorphic features, which requires a detailed study to trace the role of man as an agent of biotic weathering.

Quarries (Plate : 1) and mines carved out by human being in the Rohtas Stage of the Lower Vindhyan formation are of geomorphic importance. These quarries for extraction of limsetone are found along the left bank of the Son river in the lower slopes of the Rohtas scarps. The Tiura-Pipardih, Chunhatta, Baulia, Murli hills are important sites where the man-made weathering is conspicuous.

The mining of carbonaceous and pyritiferous beds of bijaigarh shales near Amjhor, in the different nalas, Baruanala, Ghoganala, Kaula nala, had developed significant hollows in the form of cliffs below the upper Kaimur sandstone. These mines are also peculiar example of biotic weathering .

Not only the quarries and mines but also the features due to cutting of sandstones for building purposes attest to the testimony of the man-made weathering.

The development of roads constructed by the Forest Department for the purpose of extraction of forest product and inspection presents a glaring picture of geomorphic features torn into the mass of the Plateau by man.

3.2.2.2. Run-off

Runoff consists of 'all surface water flow, both over the land slopes and in streams. Runoff may be derived directly from excessive precipitation which cannot infilterate the soil, or it may originate as the outflow of ground water along line where the water table intersects the earth's surface' (Strahler, 1971, p.415).

The study of runoff eroding and transporting rock materials, and shaping the landforms of drainage system is essential.

Seeking to escape to progressively lower levels and eventually to the sea, runoff becomes organised into drainage system, which may describe as more or less pear-shaped areas bounded of divides. Within such areas ground slopes and branching stream networks are adjusted to dispose, as efficiently as possible, of the runoff and its contained load of mineral particles, and thus to lower the land surface progressively towards the ultimate goal of the process of denudation-reduction to a nearby feature-less plain close to sea-level.

As we know, volume and rapidity of runoff is generally determined by (i) precipitation and other climatic features, (ii) nature of streams, (iii) vegetation cover, (iv) slope of the surface, (v) character of rocks and ground-water table.

As regards the Rohtas Plateau, it falls under subtropical monsoon climate where precipitation comes in the form of rainfall with an average annual of 790 mm. to 2042 mm. with alternate dry and wet periods of March to May and June to September respectively. The mean maximum temperature is about 41.78^0 C to 45.59^0 C during May and the average minimum temperature during December-January ranges from 1.00^0 C to 10.17^0 C. The monsoon generally starts by the middle of June and lasts till the end of September, the wetter months being July and August, accounting for over 80 % of the total annual rainfall. Winter rainfall in December to January is irregular and the amount is negligible. Much of the rainwater is returned directly to the atmosphere by evaporation.

Thus there is considerable variation of rainfall from season to season and from one area to another. Seasonal variation is related to the runoff variation in the Plateau.

As regards the character of the rocks and slope of the surface, the Plateau is a region of physical inequalities. The general slope of the region is in the northern direction. Extensive southern area lie above the 500 metres and northern area below 400 metres, which present a morphologic features.

The Plateau consists mainly of limestone and shale in the inner strata and sandstone in outer strata. Most of the Plateau area covered with a thin veneer of soil.

Because of torrential rain, sedimentary rocky surface and marked slopes, groundwater is limited. These character facilitated rapid runoff particularly in the steep slopes of scarps and on bare impervious rocks. The quick runoff renders the streams ephemeral.

The region is mostly covered with rich forest and scrubs jungle. The forest cover not only protect the area at places from severe soil-erosion, but also help in **interception** a process in which much of the rain at the beginning is held in droplets on the leaves and plants stems. In recent years, however, forest fires, overgrazing, shifting cultivation and mismanagement of forests have removed the vegetal cover from the surface resulting into the quick run-off specially in the rainy season.

The Rohtas Plateau gets quick and voluminous runoff during the rainy season, but it is imtermittent and ephemeral. As the distance from the water-parting increases, so also does the volume and erosive power of water flowing in the rills over the surface of the ground. On the Plateau, runoff flows down the slopes of the land more or lesss broadly distributed rilms, sheets or rills which may be designated as overland flow_In the lower and moderately sloping areas of the Plateau, **i.e.**, scarps zone, the overland flow gradually takes the place of stream flow, in which the water occupies a distinct narrow deep gorges confined by steep wall banks. During the period of rainy season water-table levels are raised and the rate of inflow into streams is increased. But after rainy season evapotranspiration begins to increase gradually. Consequently the base flow falls steadily. In summer, it becomes the minimum.

Runoff increases generally from south to north over the Plateau in response to increasing catchment. Climatic variablility is a characteristic inherent in the Plateau. The shape of the Plateau appears to be eminently suitable for frequent concentration of runoff.

The runoff of the Rohtas Plateau may be divided into two heads :

(i) **Immediate runoff** - After rain-drop the surface water makes its own path in flow from higher to lower reaches. The Rohtas Plateau has well defined rivers and streams developed by the runoff. The main rivers flowing from the Plateau area enter the plain. Excluding the Son all such rivers have their sources in the Rohtas Plateau.

The Plateau is dissected by different hill streams which serve as tributaries of the main rivers. Most of the streams remain dry except in rainy season. Only in the gorges and deep glens water remains.

(ii) Delayed runoff - It includes the springs and underground waters of the Plateau.

Spring spot the hydrological character of the Plateau. The percolation and sinkage of water during rains through the upperlying sandstones has maintained storage over the underlying impervious shale beds. Where the sandstones have been reduced and shale surface has been exposed the formation of contact springs has come into existence attributing a peculiar hydrological feature over the surface of the Plateau.

Such springs are locally known as 'Chuans' through which water flows regularly throughout whole of the year. Truly speaking, these Chuans are the only means of water for the utilisation of human beings.

The imprint of underground water is seen in the narrow glens which are locally known as 'Khoh"_These are the collapsed narrow caverns valleys but even after the rain has dropped there in the Khoh or glens the perennial streams flow showing the character of subterranean drainage. Being down there the sinkage of water from the overlying beds is possible. The Suggi Khoh, the Gupta Khoh, the Khaddar Khoh, the Makari Khoh, the Dharu Khoh etc. are signiicant deep glens into which the picture of the underground drainage in action can be observed even in the month of April and May.

Very beautiful scene of underground water is viewed in the Gupta caves in which subterranean water flow through the 'Patalganga'_(Pit rivers).

4

DRAINAGE ASPECTS

Drainage actually refers to 'the manner in which precipitation falling within an area or brought from outside is drained off' (Ahmad, 1965, p. 27). We know that slopes are ubiquitous elements of the landscape. Similarly rivers and drainage basins, which follow the slopes are also ubiquitous features of the landscape. The present humid climate, which prevails in the Rohtas Plateau, plays as important role in the character of the development of the rivers and its valleys. We shall also see that the structure and tectonic history have much influenced the development of rivers and valleys of the Plateau, and altitude, volume, slopes and profile of the Rohtas Plateau are interrelated to the rivers and basins of the Rohtas Plateau.

The Plateau is drained by many streams. The southern scarp forms the water parting between the Son and Ganges. Practicaly all the drainage of the Rohtas Plateau flows northwards to the Ganges. Son, consequently, in this part of its course, receives no tributaries of any size from the north and north-east except the river Bhawatia and Ausane. There are, however, many hill streams which originates just from the edge of the southern scarps and join the Son. Thus the Son has not been able to influence the drainage network that exists on the Plateau itself.

The Plateau is drained mainly by such streams that originate from springs existing on it and flow towards north are Karamnasha, Kohira, Karsotha, Durgawati, Dumarkhar, Dhoba etc. The Gurwat, Lawagarha, Dagia, Okhargarha, Lorsi are the main tributaries of the Karamnasha river. Among the north-flowing streams the most important is the Durgawati,. Numerous streams join it in its course taking the main stream northwards. The main tributaries of the Durgawati are Piparkhar, Goptha, Hathidah, Hasadah, Khora, Bajara, Badia, Dekura, Koel and Chormara. Another sizeable area is drained by the river Sura. Flowing in a northerly direction it maintains more or less a straight course, displaying falls and rapids as all other streams.

The volume of water in all the streams get considerably reduced in summer, smaller ones even dry up. The rains bring them to life and the numerous temporary streams and torrents and falls and rapids associated with them convert the whole Plateau into a pretty dreamland. All these rivers and streams of the Rohtas Plateau are seasonal and intermittent in character. 'They become disconnected pools of stagnant water during the cold and hot seasons, but immediately after monsoonal showers in the hills, they come down in sudden short-lived freshets. But the roaring cascades of water lasts only for a few hours, the streams subsiding again to fordability' (Ahmad, 1965, p. 30). Even during the monsoon, these streams do not maintain a regular flow which is of a very freakish nature, particularly on the marked slopes of the Plateau where the runoff is to quick to allow the streams to have a regular flow.

Most of the river valleys are V-shaped and occasionally U-shaped and sometimes apple-shaped (wide base and narrow top). The rivers are well defined by the water-partings. Sinuosity of the courses is the common feature. Waterfall and springs are other notable features.

The various aspects of drainage analysis, which will ultimately help in the understanding of the evolution of the Rohtas Plateau, may be arranged under the following heads :

4.1 DRAINAGE BASINS

The study area is a Plateau, not a basin. So it includes numerous drainage systems in both full and part courses. However, there are some nine drainage basins (Table : 4.1) which are of complete system. The Durgawati basin, although occupies a greater area of the Plateau, does not have its complete watershed within the Plateau. There are some incomplete drainage basins, too, which become a complete watershed including the area falling outside the Plateau.

The study of drainage basins has been necessiated owing to the fact that a drainage basin is considered to be the best unit for geomorphic investigation, particularly when we want to analyse the area in respect of the landscape evolution, processes involved therein and material's influence incurred there upon. In a complete drainage basin, one can see even the stages of landscape development.

TABLE : 4.1

DRAINAGE BASINS & BASIN SHAPE

Basin	Perimeter in km²	Area in km²	Max. length in km.	Max. Width in km.	Min. width	Elongation ratio	Circularity Width ratio
Durgawati	98	456	32	21.5	9	0.753	0.597
Ausane	60	205	17	18	1	0.950	0.716
Sura I	72	170	24	8	1	0.613	0.412
Sura II	53	146	20	9	3	0.682	0.653
Karsotha	55	98	69	8	3	0.162	0.407
Goptha	51	127	16	16	1.5	0.795	0.614
Koel	39	78	16	7	2	0.623	0.645
Dhoba	58	68	20	6	2	0.465	0.254
Kohira	50	68	18	13.6	1.5	0.517	0.342
Lorsi	38	34	11.5	9	3	0.572	0.296

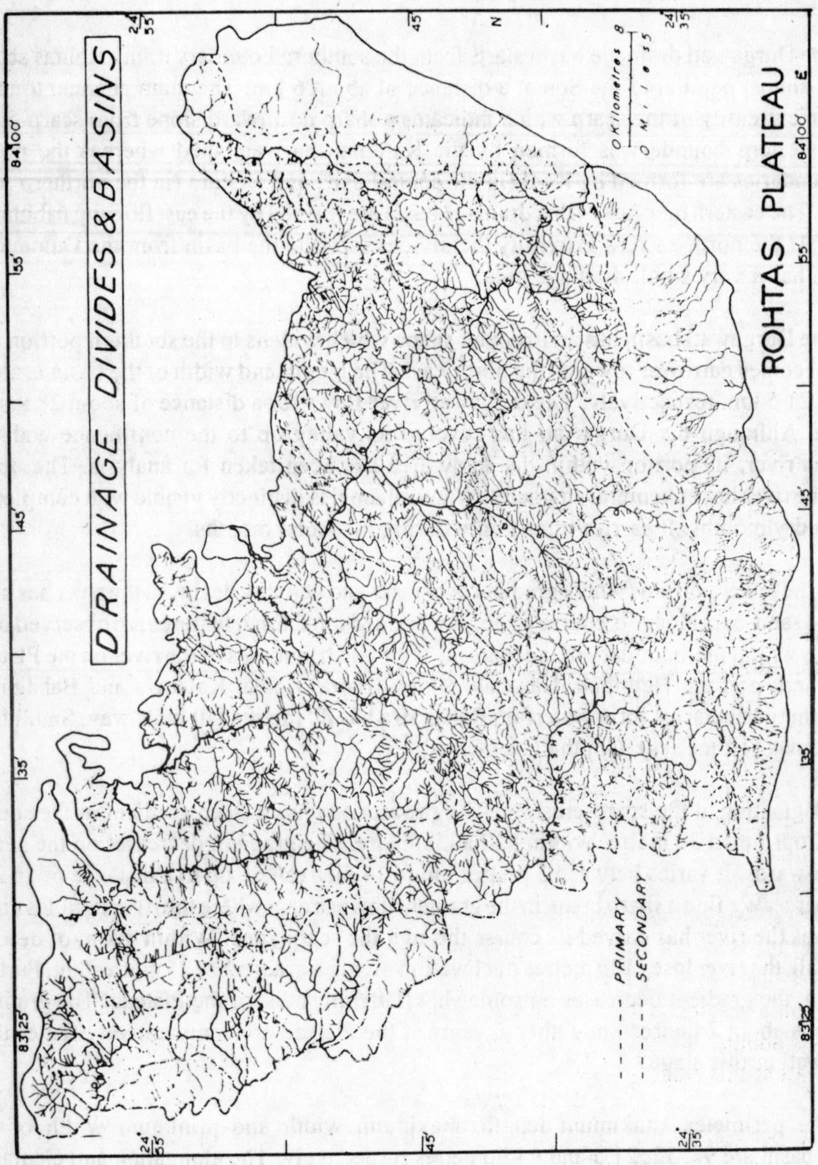
DRAINAGE DIVIDES & BASINS
ROHTAS PLATEAU
PRIMARY
SECONDARY
O Kilometres 8
O Miles 5
N
83°25
35
45
55
84°00
E
24
35

FIG. 4.1

4.1.1 DURGAWATI DRAINAGE BASIN

Comprising an area of 456 Km2, the Durgawati sub-catchment lies between Lat. 24^0 34' and 24^0 49' N. and Long. 83^0 39' and 83^0 56' E.

The Durgawati drainage basin starts from the southern boundary of the Rohtas scarps which runs almost parallel to the Son at a distance of about 6 km. The main channel touches the southern extremity of the scarp which indicates a sharp northward slope from scarp-crest. The south-western boundary is formed by the Karamnasha watershed whereas the north-western boundaries are formed by Koel (in the central part) and Dekura (in the northern part) water-shed. The eastern margins of this drainage basin are formed by the east flowing tributaries of Ausane. At the north-eastern extremity the divides separate the basin from the Dumarkhar river which has a very small drainage area.

The Durgawati basin has an irregular shape which widens in the southern portion and gradually becomes narrower towards the north-east. The length and width of the basin is about 32 km. and 21.5 km. respectively. The durgawati river flows for a distance of about 28 km. in the Plateau. Although the Durgawati has covered its valley up to the confluence with the Karamnasha river, its portion within the study area has been taken for analysis. The upper portion of the river has ephemeral character. Water channel is distinctly visible with completely dry beds in dry months. It carries runoff only during monsoon months.

In the lower portions where gradient decreases and velocity declines, the river has been able to build sand-bars along its course giving it a braided look. Such features are observed only at two points within the boundary of the Plateau. The main tributaries of the river on the Plateau in the southern part are Hathidah, Hasadah, Khora, Bajra, Badia, Kalanwa and Bahagawa. Almost all these tributaries form waterfall, Bahagawa has 80 metres fall in its way. Small falls lies as Kalanwa (10 metres), Goptha (10 metres) etc.

Originating at the scarp-crest, the Durgawati river has a steep gradient in the source region. It drops down 30 metres within half a kilometre. The gradient decreases on the gentle slope north-wards. It varies between 5 metres/km to 14 metres/km. up to a distance of 16 km. from the source. We find a sharp break in the gradient downstream where the river tumbles down a waterfall as the river has carved its course through the scarp zone. Within 4 km. of descent below the fall, the river loses 100 metres of elevation giving a gradient of 25 metres/km. Further downstream, the gradient decreases on somewhat flatter portions of the Plateau. The gradient in this area is about 7 metres/km. Other streams of the Plateau are comparatively flatter than the Durgawati at this stage.

The perimeter, maximum length, maximum width and minimum width of the Durgawati basin are 98, 32, 21.5 and 9 kilometres respectively. The elongation and cicularity ratios are 0.753 and 0.597 respectively (Table : 4.1).

The basin has a range of elevation from 100 m. to 605 m. The relative relief (Fig. 6.1) varies consideably in the basin ranging from 20 metres in the southern side, 40-60 metres in the south western side and to 350 metres in the northern side. Similarly the slope in general ranges from 2^0 to 4^0 in the upper course. But in the middle portion the slope ranges from 4^0 to 6^0. It is maximum in the scarp zone where slopes are above 12^0. The basin has a maximum slope of 16^0.

The dominant drainage pattern of the Durgawati basin is dendritic in the upper parts. But barbed pattern can be observed in some parts of the basin. The drainage texture in the upper parts of the basin is coarse (1-2/km^2) but in the middle and the lower parts moderately fine (5-6/km^2) drainage texture can be observed. The channels in the upper parts of the basin are mostly discontinued.

4.1.2 AUSANE DRAINAGE BASIN

Comprising an area of about 205 km^2, Ausane subcatchment lies between Lat. 24°37' and 24°45'45" N. and Long. 83°49'30" and 83°55'15" E. The sub-catchment drains into the Son from north-west.

The sub-catchment consists of a number of watersheds of small streams draining from west, south-west, north-east and north-west into the Ausane river. The region slopes down, in general from north-west to south-east. The sub-catchment's shape is like an old banyan tree having larger width (18 km.) than the length (17 km.). This is somewhat an uncommon shape. This is probably owing to the presence of notches and *khohs* along the tributaries coming from both the sides. It also suggests that the tributaries are guided by notches which are but the portions of subsidence over the caverns (Ch. 5.2). The main tributaries of the Ausane basin are ***Gularia nala, Barua nala, Gurdan nala, Gaighat nala, Ambakhoh nala, Kewala nala*** etc. The Ausane river flows for a distance of about 52 km.

The river rises at an altitude of 550 metres on the Plateau. The gradient is comparatively steep in its upper reaches amounting to 20 metres per km. This rate is almost constant up to 1.5 km. From 1.5 to 16 km. On its length the gradient is approximately 5 metres per km. After running up to a distance of 16 kms. the river has steep gradient on the scarp zone. The river drops down 100 metres within a distance of 2 kms. through a series of rapids and waterfalls, within the next three kms. the gradient is about 26 metres per km. Beyond 22 kms. the gradient is about 5-6 metres per km.

The perimeter, maximum length, maximum width and minimum width of the Ausane basin are 60, 17, 18 and 1 kilometres respectively. The elongation and circularity ratios are 0.950 and 0.716 respectively (Table : 4.1).

The basin has a range of elevation from 110 metres to 582 metres. The relative relief varies considerably in the basin ranging from 40 to 60 metres in the western side and 80 to 100 metres in the northern side. More than 70 % of this basin has a relative relief of more than 100 metres exceeding up to 230 to 385 metres (Fig. 6.1). Similarly the slope in general ranges from 2^0 to 4^0 on the Plateau top. But it is maximum in the scarps zone where the slopes are more than 12^0. The maximum gradient is 22^0 on the scarp zone. The slopes are below 2^0 in the alluvial plain area.

Dendritic is the most common pattern of the drainage everywhere. In the upper parts of basin the drainage texture is moderately fine (5-6/km^2). In the middle part it is fine texture (7-9/km^2). But in the alluvial plain coarse drainage texture (1-2/km^2) can be observed. The channels in the upper part of the basin are discontinued.

4.1.3. SURA I DRAINAGE BASIN

Comprising an area of 170 km2 the Sura I sub-catchment lies between Lat. 24° 43' 30" and 24° 57' 30" N. and Lon. 83° 30' and 83° 36' E. The sub-catchment drains into the Durgawati from south-west.

Sura I river has its source on the Plateau at an elevation of 475 metres. The length and width of basin is about 24 and 8 km. respectively. The Sura I river flows, for a distance of about 79.5 km. The river has more or less straight course from south to north up the scarp. From here it takes a sharp elbow turn towards the east-north-east. The basin has a shape comparable to smoking pipe with narrower area on the Plateau towards the south and the broader area in the central part. This drainage basin is bounded on the east by the basins of Sura II and Karsotha while on the west the watershed is formed by Karamnasha and Kohira drainage basins.

The gradient is roughly 10 metres per km. in the upper part up to 4 km. from the source. The gradient flatten to 4 metres per km. up to nine kilometres downward. It flattens further to about 2 metres per km. for a distance up to 18 km. This is because it flows on the gentle slope on the Plateau. The gradient starts increasing again after 35 km. The river tumbles down a waterfall 80 metres high into ***Telhar kund*** at 35.5 km. from the source. The gradient is about 10 metres to 1 km. below the fall which flattens out to 2 metres per km. in the lower reaches.

The perimeter, maximum length, maximum width and minimum width of the Sura I basin are 72, 24, 8 and 1 kilometres respectively. The elongation and circularity ratios are 0.613 and 0.413 respectively (Table : 4.1).

The basin has a range of elevation from 100 metres to 444 metres. The relative relief (Fig. 6.1) varies considerably in the basin ranging from 20 to 40 metres in the middle part. But in the scarp zone it is more than 100 metres. In the elongated deep river valleys the relative relief is more than 200 metres. Similarly the slope in general ranges from 2° to 4° in the upper part, 6° to 8° in the east and west of the basin. But it is maximum in the scarps zone where slopes are more than 12°. The maximum in the scarps zone where slopes are more than 12°. The maximum gradient is 14° on the escarpment.

Dendritic is the dominant drainage pattern in the Sura I basin. The drainage texture in the upper parts are medium. But moderately fine texture can be observed in the lower part of the basin.

4.1.4 SURA II DRAINAGE BASIN

Comprising an area of 146 km² the Sura II sub-catchment lies between Lat. 24°45' 30" and 24°56' N. and Long. 83°33' and 83° 39' E. The sub-catchment drains into the Sura I from south-east.

The length and width of the basin is about 20 and 9 km. respectively. The Sura II river flows for a distance of 53 km. The drainage basin is sandwiched between the drainage basins of Sura I on the west and Karsotha on the east. The river originates at a height of 436 metres on the west central part of the Plateau. The main channel is bow shaped while the drainage basin has a roughly triangular shape. The course is markedly sinuous in the lower reaches.

The gradient has the same increasing and decreasing pattern as Sura I. In the upper-reaches it is 10 metres to 1 km. which decreases to 4 metres to 1 km. at a distance of 12 km. It increases again to 10 metres/km. up to 19th km. At a distance of 22 km. from the source, the river makes a waterfall 130 metres high. The gradient is fairly steep below the waterfall being 20 metres/km. After 30th km. the gradient starts flattening and becomes 0.5 metres/km. in the lower reaches.

The perimeter, maximum length, maximum width and minimum width of the Sura II basin are 53, 20, 9 and 3 kilometres respectively. The elongation and circularity ratios are 0.632 and 0.653 respectively (Table : 4.1).

The basin has a range of elevation from 102 metres to 422 metres. The relative relief (Fig. 6.1) varies consideably in the basin ranging from 20 to 40 metres in the southern side, 60 to 80 metres in middle side and more than 100 metres in lower side and scarps zone. Similarly the slope in general is 2° to 4° in the upper part. But it is maximum in the scarps zone where slopes are more than 12° exceeding to 14°. But in the lower reaches the slope varies from 2° to 4°. Herring-bone drainage pattern can be observed in the Sura II basin. The drainage texture in the upper parts of the basin is medium (3-4/km²) but in the lower part it is moderately fine (5-6/km²). The channels are mostly discontinued.

4.1.5 GOPTHA DRAINAGE BASIN

Comprising an area of 127 km² the Goptha sub-catchment lies between Lat. 24°41' and 24°49' N. and Long. 83°47' and 83°55'15" E. The sub-catchment drains into the Durgawati. Goptha is a tributary of Durgawati rising on the north-eastern corner of the Plateau. The main river flows in a south-east-north-west direction while its northern tributaries flow from north to south. The main tributary of Goptha is Piparkhar. The shape of the basin is triangular. The southern tributaries have formed waterfalls along their courses. The main waterfalls are observed at ***Maspurwa nala***(Sitakund fall, 80 metres) and ***Babhandiwa nala*** (Okherian fall, 10 m.).

The gradient is steep in the source region. Within three km. from the source the river loses 65 metres of elevation giving it a gradient of about 22 metres/km. In the next 16 km. the loss in the height is 100 metres with an average gradient of 16.5 metres/km.

The river, therefore, courses through a two km. long scarp zone where the gradient is 30 metres/km. The gradient is decreases gradually downwards @ 20 metres/km. between 23.5 to 25 km. from the source and flattens out 4.5 metres/km. between 26.5 and 34 km. In the last two km. the gradient is only 2.5 metres/km.

The perimetre, maximum length, maximum width and minimum width of the Goptha basin are 51, 16, 16 and 15 kilometres respectively. The elongation and circularity ratios are 0.795 and 0.614 respectively (Table % 4.1).

The basin has a range of elevation from 135 metres to 493 metres. The relatives relief (Fig. 6.1) varies considerably in the basin ranging from 80 to 100 metres in the southern, eastern and northern side. But in the lower reaches and scarps zone relative relief is more than 100 exceeding up to 260 metres. Similarly the slope in general ranges from 6 to 8 in the upper reachcs and 4 to 6 in the northern side. It is maximum in the scarps zone where slopes are more than 12 exceeding up to 22.

The dominant drainage pattern of the Goptha basin is dendritic in the upper parts. The drainage texture in the upper parts of the basin are moderately fine (5-6/km^2) but in the lower reaches it is medium texture (3-4/km2). The channels in the upper parts of the basin are discontinued.

4.1.6. KARSOTHA DRAINAGE BASIN

Comprising an area of 98 km2 the Karsotha sub-catchment lies between Lat. 24^0 43' and 24^0 53' 30" N. and Long. 83^0 30' 30" and83^0 38'45" E. The sub-catchment drains into the Durgawati from south-west.

The length and width of basin is about 69 and 8 km. respectively. The river flows for a distance of 47 km. Karsotha has a small drainage area elongated in shape in south-west to north-east direction. The river has its source at an altitude of 480 metres close to the source of Sura I. Its boundaries are formed by the drainage basins of Sura I and Sura II on the west and koel in the east. The southern boundary is formed by the Karamnasha basin. The river has a zig-zag course in the upper parts but flows through comparatively straight channel in the central parts. Sinuosity is more pronounced in the lower reaches. The gradient in the source region is same as other rivers i.e. 10 metres/km., which decreases to 4 metres/km. where it forms a waterfall of 102 metres high. Following the pattern of other rivers gradient is steep immediately below the waterfall but flattens out the approximately 2 metres/km. in the lower parts.

The perimeter, maximum length, maximum width and minimum width of the Karsotha basin are 55, 69, 8 and 3 kilometres respectively. The elongation and circularity ratios are 0.162 and 0.407 respectively (Table : 4.1).

The basin has a range of elevation from 110 metres to 501 metres. The relative relief

(Fig. 6.1) varies considerably in the basin ranging from 40 to 60 m. in the southern side, 60 to 80 m. in the middle parts, and more than 100 m. in the northern side. Similarly the slope in general ranges from 4^0 to 6^0 in the upper part, 6^0 to 8^0 in the middle and more than 12^0 in the scarps zone exceeding up to 16^0.

The dominant drainage pattern of the Karsotha basin is dendritic in the upper parts. But radial pattern can be observed in the upper parts of the basin. The drainage texture in the upper parts of the basin is medium (3-4/km^2) but in the lower reaches it is moderately fine (5-6/km^2). The channels are discontinued in the upper parts of the basin.

4.1.7 KOHIRA DRAINAGE BASIN

Comprising an area of 88 km2 the Kohira sub-catchment lies between Lat. 24^0 05' 30" and 24^0 56' N. and Long. 83^0 24' 45" and 83^0 32' 30" E. Kohira river has its origin on the Plateau at an elevation of 400 metres. In its southern part it is called Belan river. It has two very sharp bends in its course, one in the central part where it changes its course from south-north to east-west and the other in the north-eastern part where the north-south run is again restored. The river has marked sinuosity in the northern half. The shape of basin is irregular.

The river has steep gradient in the source region, that is 60 metres/km. and this rate continues up to 2 km. from the source. It decreases downwards. Between 2-4 km. the gradient is 40 metres/km. The river flows through a gorge where the gradient is 16.5 metres/km. There is a sudden loss of steepness in gradient below the gorge area where the slope is only 2 metres/km. The result has been the creation of pronounced meanders by the river. Hood-plain, although narrow, has been produced.

The perimeter, maximum length, maximum width and minimum width of the Kohira basin are 50, 18, 13.5 and 1.5 kilometres respectively (Table : 4.1). The elongation and circularity ratios are 0.517 and 0.342 respectively (Table : 4.1).

The basin has a range of elevation from 100 to 422 metres. The relative relief (Fig. 6.1) varies considerably in the basin ranging from 80 to 100 m. in the southern side, 40 to 80 m. in the western side and 60 to 80 m. in the north-western side. It is maximum in the scarp zone where relative relief is more than 100 m. In the middle and northern side of the basin, the relative relief is more than 200 m. Similarly the slope in general ranges from 2^0 to 4^0 in the upper part of the basin, 6^0 to 8^0 in the middle and is maximum in tne scarps zone slopes are more than 12^0 exceeding up to 14^0. Rectangular and herring-bone drainage patterns are dominant in the Kohira basin.

4.1.8 KOEL DRAINAGE BASIN

Comprising an area of 78km^2 the koel sub-catchment lies between Lat. 24^0 42' 30" and 24^0 48' 30" N. Long. 83^0 37' and 83^0 44' E. The sub-catchment drains into the Durgawati from south.

The length and width of the basin is 39.5 km. Drainage basin of koel is also small. The river at an elevation of 473 metres. The basin is roughly oval shaped. The river has many sharp bends in its course especially i the upper half.

The gradient variations differ compared to other drainage basin. In the upper reaches the gradient is 10 metres/km. Which decreases to 4 metres/km. up to 19 kms. It increases to 40 metres/km. between 19 to 22 km. The river has no sharp break in slope giving rise to waterfalls like other rivers. The river has been able to cut through the steep wall modifying the slope a little. But the gradient of descent is fairly steep i.e. 40 metres/km. which continues up to lower reaches.

The perimeter, maximum length, maximum width and minimum width of the Koel basin are 39, 16, 7 and 2 kilometres respectively (Table : 4.1). The elongation and circularity ratios are 0.623 and 0.645 respectively (Table : 4.1).

The basin has a range of elevation from 103 to 422 metres. The relative relief (Fig. 6.1) varies considerably in the basin ranging from 20 to 40 m. in the upper part of the basin 60 to 80 m. in the western and northern part and more than 100 m. are in the eastern side of the basin. The slopes in general ranges from below 2^0 in the uppermost part of the basin, 2^0 to 4^0 in the middle and 6^0 to 8^0 are in the lower reaches. The channels are discontinued in the upper parts. The pattern in general is dendritic but rectangular can be observed in this basin.

4.1.9 DHOBA DRAINAGE BASIN

Comprising an area of $68km^2$ the Dhoba sub-catchment lies between Lat. 24^0 45' and $24^0$55'N. and Long. 83^0 56' 30" and 84^0 04' 15" E. The sub-catchment drains into the Durgawati.

The length and width of the basin is about 20 and 6 km. respectively. The river flows for a distance of 52 km. The Dhoba drainage basin is spread over the north-eastern corner of the study area. It is a long drainage basin but the width is small. The river has its origin at an elevation of 515 metres towards the south. The course is sinuous. The boundaries of the basin are formed by Ausane watershed in the south, and the Dumarkhar basin on the west. Its eastern boundary runs along the escarpments.

The river has a much higher gradient in the source region i.e. 20 metres/km. to 40 metres/km. up to 6 km. from the source. It decreases to 3 metres/km. up to6 km. from the source. It decreases to 3 metres/km upto 15 km. but increases again downstream. The gradient varies 20 metres to 40 metres/km. between 15th to 17th km. In the next 20 km. the gradient varies between 2 metres/km. to 10 metres/km. At 44 km. from the source the river descends through a waterfall (Dhuadhar) 30 metres high. The gradient below the fall is 30 metres/km. which flattens out to 2.5 metres/km. in the lower parts.

The Perimeter, maximum length, maximum width and minimum width of the Dhoba basin are 58, 20, 6 and 2 kilometres respectively (Table : 4.1). The elongation and circularity ratios are 0.465 and 0.254 respectively (Table : 4.1).

The basin has a range of elevation from 122 to 523 metres. The relative relief (Fig. 6.1) varies considerably in the basin ranging from 60 to 80 m. in the southern side of the basin. It is more than 100 m. in the middle part of the basin. But in the northern side the relative relief varies between 20 to 40 m. Similarly the slope in general ranges from 4^0 to 6^0 in the upper parts, 6^0 to 8^0 in the middle and more than 12^0 in the scarps zone. The maximum gradient is 20^0.

The dominant drainage pattern of the Dhoba basin is dendritic in all the parts. But radial pattern can be observed in the upper parts of the basin. The drainage texture in the upper parts of the basin is fine (>9/km2), but in the lower reaches it is moderately fine (5-6/km^2). The channels are discontinued in the upper reaches of the basin.

4.1.10 LORSI DRAINAGE BASIN

Comprising an area of 34km^2 the Lorsi sub-catchment lies between Lat. 24^0 49' and 24^0 54' N. and Long. 83^0 22' and 83^0 26' E. The sub-catchment drains into the Karamnasha from east.

The length and width of basin is about 11.5 and 9 km. respectively. The drainage basin of this river is smallest. Its shape resembles with Australia. The river has its source on the Plateau at an elevation of 335 metres. This river forms the northernmost tributary of Karamnasha falling within the study area. Its eastern boundaries are formed by the basin of Kohira river while the southern boundary is formed by the Karamnasha divide.

The gradient is steep 4 km. from the source where it is 40 metres/km. which flattens to 3 metres/km. between 4 km. to 11 km. of the length. The gradient again increases to 20 metres/km. before it joins the Karambasha.

The perimeter, maximum length, maximum width and minimum width of the Lorsi basin are 38, 11.5, 9 and 3 kilometres respectively (Table : 4.1). The elongation and circularity ratios are 0.572 and 0.296 respectively (Table : 4.1).

The basin has a range of elevation from 92 to 330 metres. The relative relief (FIg. 6.1) varies considerably in the basin ranging from 80 to 100 m. in the south eastern side, 60 to 80 m. in the northern, and 80 to 100 m. in the northern side. The relative relief in the western and north western side is more than 100 m. Similarly the slope in general ranges from 2^0 to 4^0 in the upper parts of the basin, 6^0 to 8^0 in the middle and more than 12^0 in the scarps zone.

The dominant drainage pattern of the Lorsi basin is dendritic. The drainage texture in this basin is medium (3-4/km^2).

The study of all the ten drainage basins discussed above reveal the following points of geomorphic significance.

i)Almost all the drainage basins have uncommon shape particularly in regard to the length-breadth ratio. They have larger width than the length. The Kohira, Goptha, Durgawati, Ausane river basins are the clear-cut examples (Fig. 4.1). The Sura I, Sura II, Dhoba, Koel and Karsotha river basins seem to be elongated in shape but they are not actually separate river basins; and if Sura I and Sura II are considered together and the Karsotha and Koel basins are tagged with the Durgawati basin the width of the basin becomes larger than the length. This is a common feature with all the river basins over the Rohtas Plateau. This is a deviation from the general rule that a river has narrow catchment at the surface, it becomes wider in the middle and again it becomes over a surface of homogeneous structure. The upper surface of the Rohtas Plateau has no doubt homogeneity in structure, that is the Vindhyan sandstone throughout the Plateau, but underlying beds of limestone and shale left great impact on the development of streams and the shape of the basins. The underlying limestone beds have facilitated the developments of intricate patterns of caverns and after the subsidence of the roofs the river valleys have been formed haphazardly and owing to these subsided valleys the shape of the basins have become uncommon and haphazard.

ii) The descriptions of the river basins highlight one common feature in all the basins and that is related to the gradient of main channels where we observe that almost every channel has moderate gradient in the upper course, it sometimes increases in the middle, it suddenly drops down with the steep slopes of waterfall followed by a plung - pool at the base of the waterfall and flattened lower courses. The presence of waterfalls almsot along all the rivers and the valleys below the waterfalls with very steep side-walls (sometimes more than 90^0) do suggest that the lower portions of all these river basins have been subject to subsidence, certainly along the limestone caverns.

iii) In all the basins we have noted that in the upper reaches main streams are discontinued. This may be due to percolation of water into the underlying limestone beds through cracks and joints. This may suggest that there is continued process of development of karst features.

iv) Although all the ten drainage basins have been demarcated with their full courses from source to mouth the trunk streams of all these basins meet with either the Durgawati or Karamnasha or the Son. Therefore, they do not attain all the three stages - youth, mature and old. These basins comprise virtually the portions of youth and late youth stages. The high relative relief in most of the basins and the high gradient along the most of the streams are in favour of this deduction.

v) In the table : 4.1 circularity ratio * and elongation ratio** of all the basins have been calculated. The circularity ratio ranges from 0.254 for Dhoba basin to 0.716 for the Ausane

---

* Circularity Ratio (CQ) = $\frac{\text{Area of the Basin}}{\text{Are of the circle with same perimeter of the basin}}$

or $C = \frac{4\pi A}{P2}$ where, A = Area of the Basin, P = Basin Perimeter

** Elongation Ratio (E) = $\frac{\text{Diameter of the Circle with same area as the Basin}}{\text{Basin length}}$

or $E = \frac{2/A/\pi}{L}$ where, A = Basin Area, L = Basin Length basin. Similarly the elongation ratio is 0.162 as minimum for the Karsotha basin and 0.950 as maximum for the Ausane basin. Miller found that 'circularity ratio Rc, defined as the ratio of basin area Au to the area of a circle Ac having the same perimeter as the basin. He found that circularity ratio remained remarkably uniform in the range 0.6 to 0.7 for first - and second - order basins in homogeneous shales and dolomites, indicating the tendency of small drainage basins in homogeneous geologic materials to preserve geometrical similarity. By contrast, first - and second - order basins situated on the flanks of moderately dipping quartzite strata of Clinch Mountain, Virginia, were strongly elongated and had circularity ratios of between 0.4 and 0.5 generally'. (Strahler, 1964, p. 51). Similarly the elongation ratio' runs between 0.6 and 1.0 over a wide variety of climatic and geologic type. Values near to 1.0 are typical of regions of very low relief, whereas values in the range 0.6 to 0.8 are generally associated with strong relief and steep ground slopes' (Strahler, 1964, p. 51). If we examine the data of circularity and elongation ratios calculated for the basins of the Rohtas Plateau in the light of the deductions made by Miller and Schumm, the Ausane, Sura II, Goptha and the Koel river basins seem to be of homogeneous geologic materials and the same basin along with the Durgawati and Sura I seem to be associated with strong relief and steep ground slopes. So far as the second deduction is concerned it is more or less true. But as regards the first deduction it does not apply to this region because the lower reaches of almost all the basins consist of limestone and shale whereas the upper reaches consist of the Vindhyan sandstone. So here is a deviation from the rule. It is because of the fact that the catchment areas of the basins have not been determined only by the simple denudational process of fluvial actions but there has been a vital role of the underground drainage resulting into the origin of a net of cavers and ultimately their collapse. So, the basin's shape are not keeping pace with the above noted deductions. It must be noted, therfore, that the methods cannot be applied to all the areas blindly. The basin's shapes of the Rohtas Plateau are largely determined by the underlying caverns.

4.2 DRAINAGE PATTERNS

The drainage maps (Fig. 4.1) of the Rohtas Plateau reveal the hierarchy of streams from their sources to their master streams. The patterns are apparently guided by factors like joints, lithology, relief and epeirogenetic movements over each of terrain. Therefore, the study of drainage pattern of the Rohtas Plateau will be one of the most practical approaches to an understanding of structural and lithological control of landform evolution.

The most common patterns of drainage in the Rohtas Plateau are dendritic, pinnate and herring-bone. Besides these, rectangular, parallel, radial, barbed or fish-hook and other patterns are also observed in different localities. The dominant drainage pattern of the Plateau is dendritic. This pattern can be observed in many parts of the Plateau **e.g.**, in the upper reaches of the Durgawati, Ausane, Karamnasha, Koel **etc.** river basins. These have evolved on account of the homogeneous structure of the Kaimur sandstone forming the upper layers of the Plateau. The pattern develops in areas of minimum structural control and 'are characterized by irregular branching of tributary streams in many directions and at almost any angle, although usually at considerably less than a right angle' (Thornbury, 1962, pp. 120-21). 'This kind of pattern is interpreted as signifying rocks of unifrom resistance to erosion. This indicates thick horizontal sediments or a massive, homogeneous crystalline bedrocks' (Morisawa, 1985, p. 159). The upper sedimentary strata of the Plateau are completely free from folds, faults and other structural determent except for a gentle downwarp towards the north.. Widespread dendritic character of the drainage lines is, therefore, the result.

Closely related to the dendritic pattern, the pinnate pattern is very conspicuous over many areas of the Plateau. It is a special case of dendritic pattern in which the tributaries to the main stream are sub-parallel and join it at acute angles. They are conspicuous in many parts of the Plateau especially in the drainage basins of Koel, Karsotha, Ausane which is shown in Figs. 4.12, 4.14 & 4.15. 'A pinnate pattern is considered a subtype of dendritic drainage, it is characterized by featherlike extension of many closely spaced, short tributariess. This pattern is indicative of an area underlain by easily eroded sediments and underground sapping' (Morisawa, 1985, p. 160). The Plateau, as we know, is underlain by limestone and shale beds. So, the development of this pattern is expected.

When we concentrate our attention on individual stream, herring-bone pattern, **i.e.**, a main relatively straight stream joined by roughly parallel tributaries at right angles, is strikingly conspicuous over many areas. The development of this pattern is controlled by systems of joints occurring in the drainage area. In this pattern, 'the tributary streams are rarely allowed to bend downstream to form acute-angled confluence' (Ahmad, 1985, p. 168). The tributaries join the trunk stream at right angles from both sides giving it a characteristic herring-bone appearance. This pattern is found in the basin of Okhargarha which is a tributary of Gurwat river (Fig. 4.1). This pattern is conspicuous also in the upper part of Kohira river (known also as Belan) and in the north-eastern part of Sura I basin. This is due to presence of small parallel streams flowing down on the high gradient slopes on the walls of notches and subsided valleys.

Drainage lines radiate on all sides from a single dome or a conical hill giving it a radial pattern. This is sometimes found at the junction of two drainage basins where part of the drainage goes to one basin and the part to another. On the Plateau the radial pattern is found at the junction of Sura I, Karsotha and Karamnasha drainage basins. There are numerous other examples. They can be seen at the junction of Karamnasha and Durgawati also. They have developed on isolated domal areas within the basin at several places, **e.g.** in the north-western part of Karamnasha basin and ***Arjuna Pahar*** near Banjari on the divides of Ausane river.In rectangular drainage patterns, both the main streams and its tributaries display right angle bends. They have developed along well jointed portions of the Plateau. Example are not many in the area of study. The north-western tributaries of Kohira, however, exhibit this pattern.

Barbed pattern which is also known as fish-hook pattern, has developed in certain areas of the the Plateau. 'Most barbed patterns are the result of stream piracy which has effected a reversal of the drainage of part of a separate river system' (Thornbury, 1962, p. 123). This pattern has developed in Karamnasha basin near village Garale (24^0 35' 15" N. and 83^0 35' 05" E.) and Durgawati basin north of the village Nawakadih (24^0 35'30" and 83^0 45' 45" E.) 'Parallel patterns are usually found where there are pronounced slope or structural controls which lead to regular spacing of parallel or near-parallel streams' (Thornbury, 1962, pp. 125-126). The development of parallel pattern is conspicuous in the southern part of the Plateau below the scarps. The streams rising near scarp-crest flow in a steep gradient to join the river Son. Many first order streams fail to join the Son on the eastern face of plateau. But the drainage lines are markedly parallel on the scarp.Thus the arrangement of streams expresses various patterns which have developed in different parts of the area in relation to the structural and physiographic control. Similarly many individual streams of the area outline their individual patterns in the form of straight, meandering, rectangular and braided nature of their channels.

Treatment of individual streams and their arrangement into some specific patterns express many relevant morphological facts. The study further aims to arrive at specific geomorphological conclusions for which the evolution of drainage system of the region will be effective only when the hydrographic network is analysed.

As has been noted earlier, the Plateau is free from folds in upper portions. The area consisted of Rewa and Bhander series before the Tertiary period which were completely eroded away exposing the Kaimur sandstone over which the present drainage lines have developed. The Kaimur sandstones were tilted slightly to the north giving the drainage lines a northward flow out the Plateau. The sandstones are of homogeneous structure throughout the Plateau but innumerable joints and fractures criss-crosses the Plateau surface which have provided probably initial orientation to stream.The primary joints and fractures which provided lines of weakness in the general flow pattern have turned in youthful consequent streams assoicated with waterfalls and rapids. The secondary joints have given rise to subsequent streams which join the master streams at angles decided by joint directions which sometimes produce acute angle confluence or right angle confluence.

The sinuosity exhibits structural control, mainly the control of joints. Meandering is not well marked even in the lower reaches suggesting to apparent youthfulness of drainage organization. It must, however, be noted that main trunks are the result of collapse of roofs over the caverns, and thus the stream of the Rohtas Plateau have adjusted with the nature and phenomena associated with the underlying rocks particularly limestones. The control of general gradient on the streams, however, cannot be denied.

4.3 HYDROGRAPHIC NETWORK

The growth of drainage basins is marked by a growing number of stream channels which unite to form confluence points. The distribution of source-heads and confluence points (Figs. 4.2 & 4.3) expresses the phases of drainage evolution and clearly depicts the nature of land and influence of structure.The maps (Figs. 4.2 & 4.3) very clearly show that the maximum concentration of source- heads and confluence points occur in the area lying between almost 83°30' Longitude and 84° 00' Longitude and 24° 40' north and the northern boundary of the Plateau. The maximum concentration is in the Dhoba, Durgawati , Sura I, Sura II river basins. The remaining areas have less concentration, and the minimum is in the outside the scarp zone and the north-western portion of the Plateau.

For detailed study, the number of source-heads and confluences have been counted in one square kilometres grids of the entire Plateau from 1:50000 scale topographical sheets and their respective figures have been inserted on the 1cm. = 1 km. map of the country to draw isopleth maps (Figs. 4.2 & 4.3). These maps bring out clearly the density of source-heads and confluences and pattern of different categories. Again these figures have been categorised into different density zones and their respective areas have been shown by histograms attached there in (Figs. 4.2.2 and 4.3.3).

Table 4.2 gives the six categories of density with number of source-heads, total units in square kilometres and their percentage. Five major density zones of source-heads, viz., High, moderately high, medium, low and cypher (nought or zero), cover 4.58, 10.85, 25.00, 35.34 and 24.23 per cent respectively in the total area of the Rohtas Plateau.

TABLE : 4.2
DENSITY OF SOURCE HEADS

No. of cat.	No. of source heads per unit	Total area in square kms.	% to the total area	Cumulative freq.%	Density zones	% to density zones
I	0	567	24.23	24.23	Cypher	24.23
II	1-2	827	35.34	59.57	Low	35.34
III	3-4	585	25.00	84.57	Medium	25.00
IV	5-6	254	10.85	95.42	Moderately high	10.85
V	7-8	94	4.02	99.44	High	4.58
VI	9	13	0.56	100.00		
Total		2340	100.00			100.00

TABLE : 4.3
DENSITY OF CONFLUENCE POINTS

No. of cat.	No. of C.P. per unit	Total area in km²	% to the total area	Cumulative freq. %	Density zones	% of Density zones
I	0	620	26.50	26.50	Cypher	26.50
II	1-2	908	38.80	65.30	Low	38.80
III	3-4	552	23.59	88.89	Medium	23.59
IV	5-6	186	7.95	96.34	Moderately high	7.95
V	7-8	54	2.31	99.15	High	3.16
VI	9	20	0.85	100.00		
Total		2340	100.00			100.00

Similarly Table 4.3 shows the six categories of density with number of confluences, total units in square kilometres and their percentages.

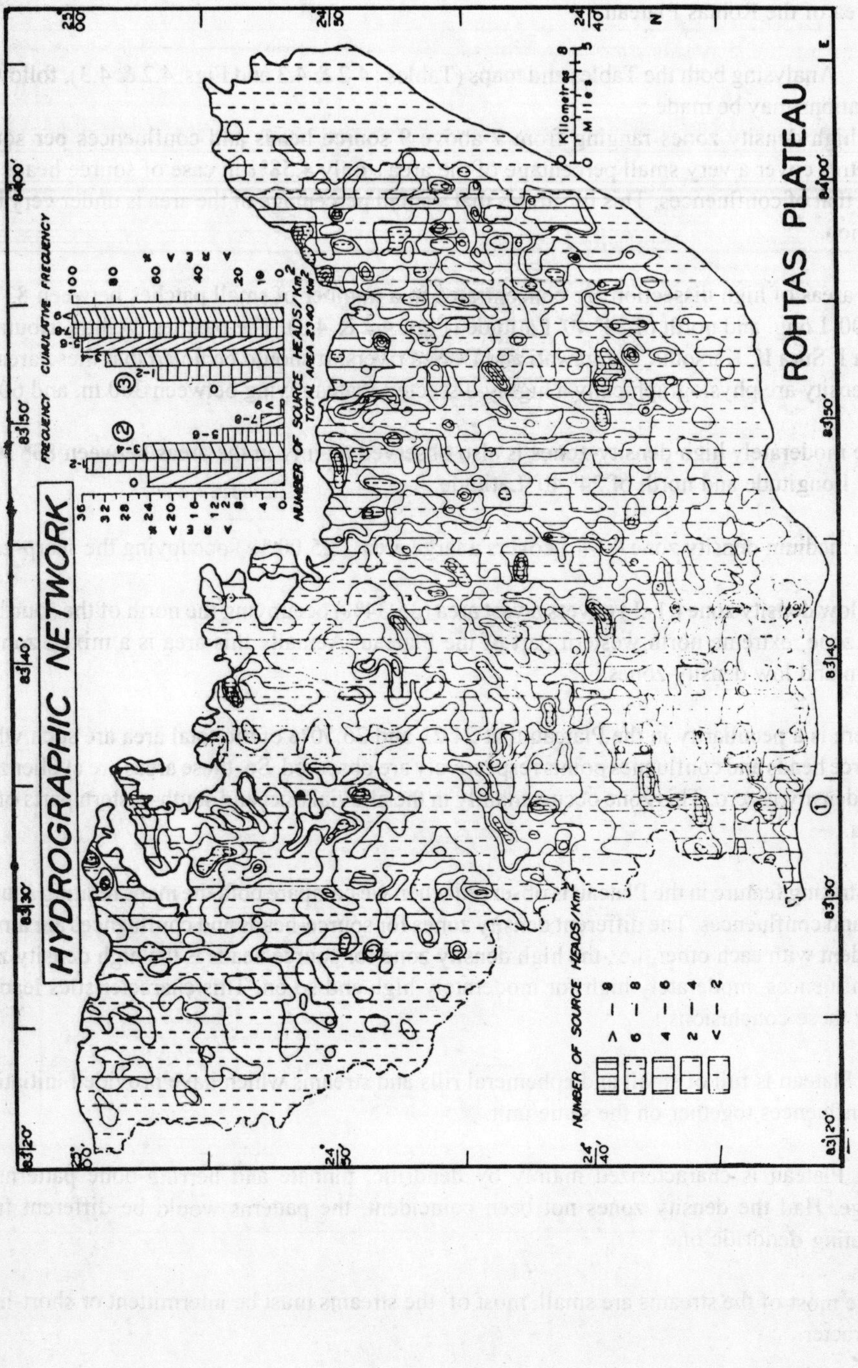
HYDROGRAPHIC NEWORK
ROHTAS PLATEAU
NUMBER OF SOURCE HEADS/Km²
TOTAL AREA 2340 Km²
FREQUENCY
CUMULATIVE FREQUENCY
Kilometre
Miles
N
E

FIG. 4.2

In this case also five major density zones clearly crop out viz., high, moderately high, medium, low and cypher covering 3.16, 7.95, 23.59, 38.80 and 26.50 per cent respectively to the total area of the Rohtas Plateau.

Analysing both the Tables and maps (Tables : 4.2 & 4.3 and Figs. 4.2 & 4.3), following observations may be made :

i. The high density zones ranging from 7 above 9 source heads and confluences per square kilometres cover a very small percentage of the area - only 4.58% in case of source heads and 3.16 in that of confluences. This illustrates that a small percentage of the area is under very high dissection.

ii. The areas of high dissection are concentrated in a number of small patches between 83^0 40' to 84^0 00' Long. and north of 24^0 40' Latitude (Figs. 4.2 & 4.3). The patches include tributaries of Sura I, Sura II, Karsotha, Durgawati and Dhoba rivers. It should be noted that these areas of high density are physiographically a highly dissected Plateau lying between 300 m. and 600m.

iii. The moderately high density-zones is also observed mainly in the areas between 83^0 30' to 84^0 00' Longitude and north of 24^0 40' Latitude.

iv. The medium density zone (3-4) covers a large area (25.00 %) occupying the scarp-zone.

v. The low density zone (1-2) covers a vast area (35.34%) occupying the north of the sourthern scarps zone, extreme north western part of the Plateau. Actually this area is a mixed zone of medium and low density zones.

vi. There is a peculiarity in the Plateau that 24.23 and 26.50% of the total area are such where no source heads and confluencs points respectively are observed. So, these areas are cypher zone where density is zero. This zone occurs mainly in the alluvial tract and south-eastern parts of the Plateau.

vii. A striking feature in the Plateau is observed when we compare both the maps relating source-heads and confluences. The different density zones for source-heads and confluences are almost coincident with each other, i.e., the high density zone for source heads is the high density zone for confluences, moderately high for moderately high and so on. This characteristics leads to mainly these conclusions :

1. The Plateau is full of small and ephemeral rills and streams which have provided initiations and confluences together on the same unit.

2. The Plateau is characterized mainly by dendritic, pinnate and herring-bone patterns of drainage. Had the density zones not been coincident, the patterns would be different from dominating dendritic one.

3. Since most of the streams are small, most of the streams must be intermittent or short-lived in character.

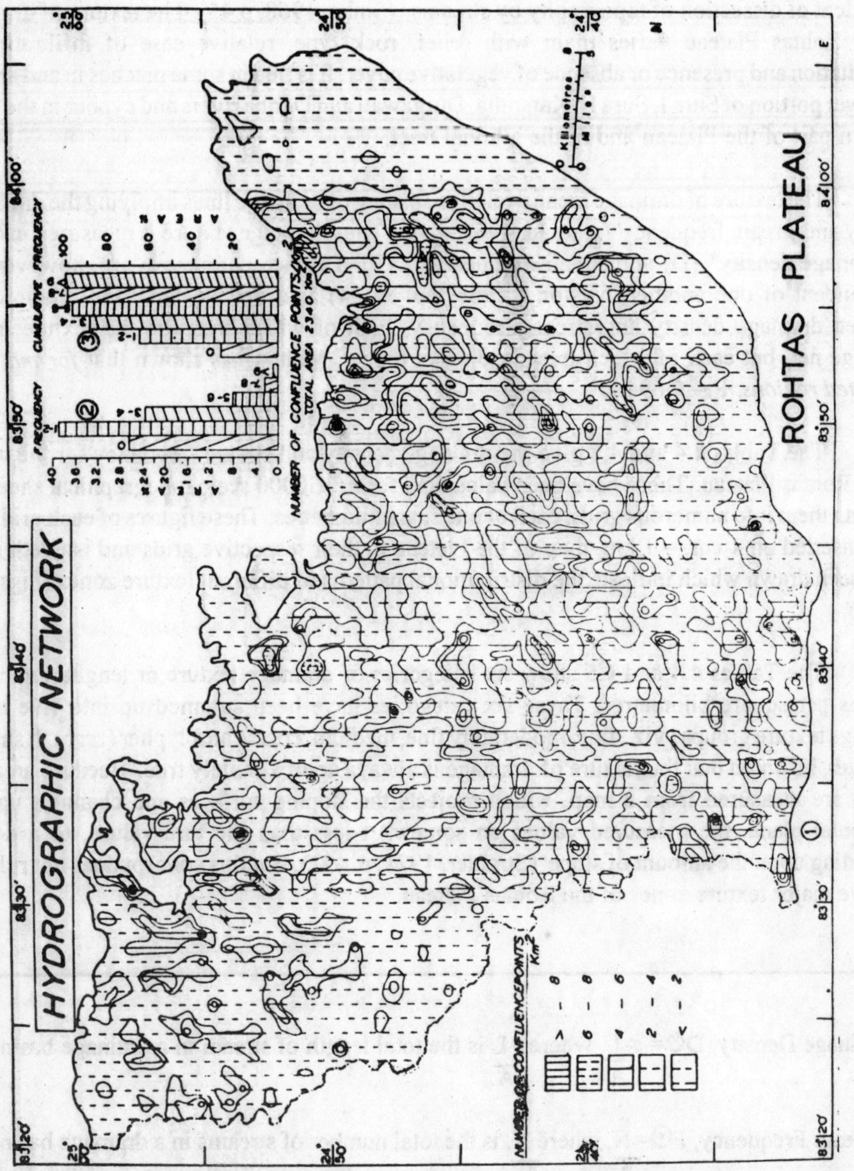
HYDROGRAPHIC NETWORK
ROHTAS PLATEAU
NUMBER OF CONFLUENCE POINTS
TOTAL AREA 2340 Km²
FREQUENCY
CUMULATIVE FREQUENCY
AREA
NUMBER OF CONFLUENCE POINTS Km²
Kilometres
Miles
83°20'
83°30'
83°40'
83°50'
84°00'

FIG. 4.3

4.4 DRAINAGE TEXTURE

Drainage can be graded as coarse, medium, fine and ultrafine depending on the degree and extent of dissection of topography by streams ' (Sinha, 1968, p.45). The texture of drainage in the Rohtas Plateau varies main with relief, rock type, relative ease of infiltration of precipitation and presence or absence of vegetative cover. It is fine in some patches in and around the lower portion of Sura I, Sura II, Karsotha, Durgawati and Dhoba rivers and cypher in the north western part of the Plateau and in the alluvial tract.

The texture of drainage means' relative spacing of drainage lines implying the drainage-density and stream frequency' (Ahmad, 1965, p.19). Thus drainage texture is measured in terms of drainage density* (D) and stream frequency** (F). The two parametres are, however, not independent of one another. Melton (1966, pp. 35-54) analysed in detail the relationships between drainage density and stream frequency, both of which measure the texture of the drainage net, but each of which treats a distinct aspect. Melton has shown that *for maturely dissected regions,* $F = 0.694\ D^2$.

The Tables 4.4 and 4.5 give the drainage density and stream frequency for the whole of the Rohtas Plateau. These have been calculated form 1:50000 scale topographical sheets by dividing them into numerous grids, each of one sqare kilometres. These figures of each grid have been inserted on 1 cm. = 1 km. map of the Plateau in their respective grids and isopleth maps have been drawn which indicate the distributional patterns of different texture zones (Figs. 4.4. & 4.5).

The Tables 4.4 and 4.5 show six categories of drainage texture or lenght/number of streams per square kilometres. These six categories have been summed up into five major drainage texture groups ; viz., fine, moderately fine, medium, coarse and cypher (zero). It should, however, be noted that the texture of drainage is alwaya approximately true. 'Because area and length are measured from a map, which projects the sloping surfaces and channels upon a horizontal plane, the measured values on somewhat less than the ture values, to a degree depending upon the amount of slope' (Strahler, 1971, p. 491). But this concept does not rule out the five major texture zones of the Rohtas Plateau.

---

* Drainage Density, $D^{\otimes} = \frac{\Sigma L}{A}$, where $\frac{\Sigma L}{A}$ is the total length of stream in a drainage basin area A.

** Stream Frequency, $F^{\otimes} = \frac{N}{A}$, where N, is the total number of streams in a drainage basin area A.

⊗ Introduced into American Hydrology and literature by Horton (1945), in 'Erosional Development of streams and their drainage basins : hydrological approach to quantiative morphology, Bull. Geol. Soc. Am., 56, pp. 275-370.

DRAINAGE DENSITY
ROHTAS PLATEAU
LENGTH OF STREAMS /Km2
KILOMETRES
LENGTH OF STREAMS Km2
FREQUENCY
CUMULATIVE FREQUENCY
AREA %
Kilometres
Miles

FIG. 4.4

TABLE : 4.4
DRAINAGE DENSITY*

(Lenght of streams in kilometre per sq. kilometre)

No. of cat.	Length of streams in km²	FREQUENCY Km²	%	cumulative %	Density zones	% of the density zones
I	0	228	9.74	9.74	Cypher	9.74
II	.5-1.0	425	18.16	27.90	Coarse	18.16
III	1.5-2.0	617	26.37	54.27	Medium	26.37
IV	2.5-3.0	739	31.58	85.85	Moderately fine	31.58
V	3.5-4.0	259	11.07	96.92	Fine	14.15
VI	4.0	72	3.08	100.00		
Total		2340	100.00			100.00

TABLE : 4.5
STREAM FREQUENCY

(Number of streams per sq. kilometres)

No. of cat.	No. of streams per unit	FREQUENCY km²	%	cumulative %	Texture zones	% of the texture zones
I	0	122	5.21	5.21	Cypher	5.21
II	1-2	403	17.22	22.43	Coarse	17.22
III	3-4	678	28.97	51.40	Medium	28.97
IV	5-6	685	29.28	80.68	Moderately fine	
V	7-8	341	14.57	95.25	Fine	17.32
VI	9	111	2.75	100.00		
	Total	2340	100.00			100.00

4.5.1 THE FINE DRAINAGE TEXTURE

It includes the patches of the highest drainage density (3.5 to 6 kilometres / km²) and highest stream frequency (7 to 13 / km²) covering areas of 331 km² (14.15 %) and 452 km² (17.32 %) respectively. Most of the areas lying under this texture have drainage density of 3.5 - 4.0 kilometres / km² and stream frequency of 7-8/km² as is clear from the above tables.

The fine texture is found mostly in the lower part of the all river basins. It includes the drainage basins of Sura I, Sura II, Karsotha, Durgawati, Dhoba and Ausane.

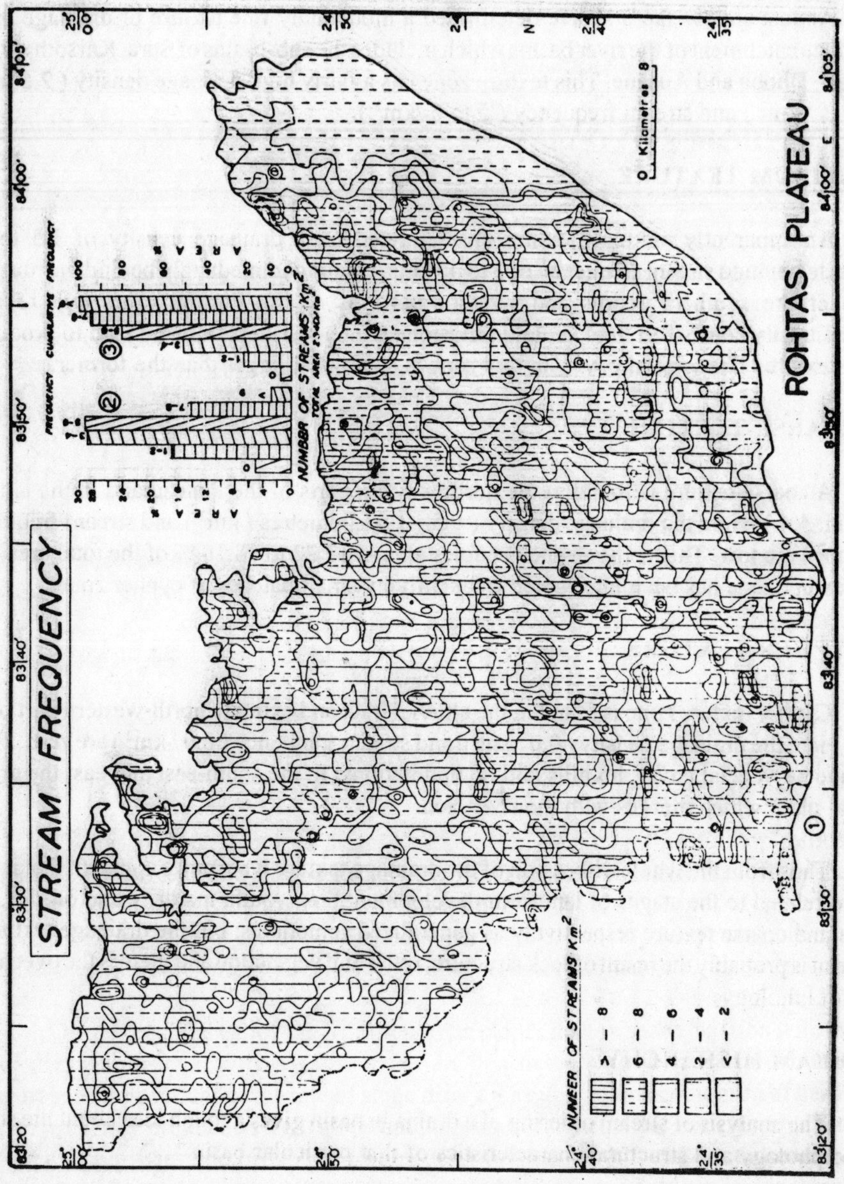

FIG. 4.5

These areas include mostly the notches and ***Khohs*** and characterized with sparse vegetative cover.

4.5.2. MODERATELY FINE TEXTURE

Almost similar factors have determined a moderately fine texture of drainage in the intermediate catchment of the river basins which includes the sub-basins of Sura, Karsotha, Koel, Durgawati, Dhoba and Ausane. This texture zone has a fairly high drainage density (2.5 to 3.0 kilometres / km^2) and stream frequency (5 to 6 /km^2).

4.5.3. MEDIUM TEXTURE

An apparently medium texture characterized by a drainage density of 1.5 to 2.0 kilometres / km^2 and stream frequency of 3 to 4 / kms. is found immediately behind the northern portions of the southern scarps and western portions of the Plateau. This texture zone characteristically consists of hard and old sandstone which resist erosion and tend to produce a medium texture. The vegetal cover, in this zone is somewhat larger than the former.

4.5.4. COARSE TEXTURE

A coarse texture is found in the north western parts of the Plateau and in the alluvial tract (Fig. 4.4). Here the drainage density (0.5 - 1.0 kilometres / km^2) and stream frequency (1-2 / km^2) are low. This texture zone occupies almost 17.22 to 18.16 % of the total area. The coarse texture zone has been interrupted frequently by small patches of cypher zone.

4.5.5 CYPHER TEXTURE

Cypher texture zone is found in the alluvial tract and extreme north-western part of the Plateau. Here the drainage density (0.0 / km^2) and stream frequency (0.0 / km^2) are zero. These areas have no vegetal cover, no hills and no undulations. In the south-east and east the area is a alluvial plain which merges with the river Son.

Thus from the whole discussion of the drainage texture it becomes apparent that it does not seem related to the stages of landform development, **i.e.**, youth, maturity and old for fine, medium and coarse texture respectively as generalized sometimes; but the drainage texture of the Plateau is probably the result of rock structure, relief and vegetation and there is the overriding control of lithology.

4.5 STREAM HIERARCHY

The analysis of stream ordering of a drainage basin gives a rough idea about the shape, size, morphology and structural characteristics of that particular basin.

The data for different basins facilitate a comparative analysis of all these geomorphological components. On the basis of the study of stream ordering of a number of drainage basins under various conditions, these are some deductions; and these deductions make it possible to infer condition pertaining to an unknown basin even if opportunity for direct observation is not available.

There are some 6 - 7 scheme of stream ordering presented by geologists and geomorpholgists working on properties of drainage basins. Important among them are the schemes suggested by Horton (1945), Strahler (1952), Scheidegger (1965) and Shreve (1967) although suggestion by Woldenberg (1966) deserve a considerable attention.

It is noted that the first scheme of stream ordering was suggested by Gravelius (1914), which received little attention. It was the Horton's scheme (1945) that became popular in geomorphology. But its modification by Strahler (1952) was widely accepted because of later's simplicity and former's complicacy. In Horton's scheme there are (i) two stages of stream ordering, (ii) subjective decision and (iii) classification of unbranched tributaries into different orders (Gregory & Walling, 1976, p. 41 and Doornkamp & King, 1971, pp. 4 - 5). No doubt, these suggestions have sound logical base, and they can be regarded as valid propositions if some deductions are made by applying the methods of Scheidegger and Shreve to a variety of area under different conditions. Unless it is done the Strahler's scheme cannot be set aside. The author has, therefore, followed the scheme suggested by Strahler.

The present work analyses ten drainage basins. For the purpose basins have been demarcated and a quantitative analysis of their drainage composition has been made. The basins have been traced from 1:50000 topographical maps. The network has been sub-divided into individual lengths of channel or channel segements, according to a hierarchy of orders of magnitude on Strahler's scheme. The numbers and lengths of segments of each order were then counted and measured to yield the figures in Tables 4.6.1., 4.6.2, 4.6.3, 4.6.4, 4.6.5, 4.6.6. 4.6.7, 4.6.8, 4.6.9., and 4.6.10. With the help of these figures, logarithms of the number (Nu), Length (ΣLu) and cumulative mean length ($\bar{L}$u) of the streams* are plotted against order (u) to show a linear relationship (Figs. 4.6 and 4.7) of the drainage network of the different order drainage basins of the Rohtas Plateau.

A study of the tables (4.6.1, 4.6.2, 4.6.3, 4.6.4, 4.6.5. 4.6.6, 4.6.7, 4.6.8, 4.6.9, 4.6.10), maps (Figs. 4.6 and 4.7) and logarithms leads to the following findings conclusions :

---

* VARIABLE	SYMBOL
i) Stream Order	u
ii) Number of streams of order u	Nu
iii) Total No, of streams within basin order u	(ΣN) u
iv) Total length of stream of order u	Lu
v) Mean length of streams of order u	$\bar{L}u = Lu / Nu$
vi) Total stream length within basin of order u	$(\Sigma L)\, u = \underline{L_1} + L_2$ --- L_4
vii) Stream length ratio	$Rl = \bar{L}u / \bar{L}u - 1$

a) In a bsin of homogeneous structure, the bifurcation ratio (Rb) tends to remain between 3 and 5. ' Values of bifurcation ratio between 3 and 5 are characteristic of natural stream system' (Strahler, 1971, p. 484). The Rohtas Plateau is structurally composed of almost horizontal strata of limestone, shale and sandstone, and there is a thick capping of the Vindhyan sandstone, about 150 to 300 metres. There is no sturctural variation over the Plateau top (Ch. 2.5). Therefore, theoretically Rb should remain between 3 and 5. But it has not been maintained in all the basins under study. Only three drainage basins - Goptha, Kohira and the Lorsi basin conform to the findings of Strahler (Tables 4.6.1, 4.6.2 and 4.6.3).

TABLE : 4.6.1
BASIN MORPHOMETRY OF GOPTHA

Stream Order	No. of streams or segements	Bifurcation Ratio	Total Length	Mean Length of segments in kms.	Cumulative mean length in km.	Length Ratio
u	Nu	Rb	kms	Lu	$\bar{Lu}$	RL
First	419		394	0.94	0.94	
Second	93	4.51	131	1.41	2.35	1.50
Third	24	3.88	56	2.33	4.68	1.65
Fourth	5	4.80	48	9.60	14.28	4.12
Fifth	2	2.50	11	5.50	19.78	0.57
Sixth	1	2.00	15	15.00	34.78	2.73
		17.69	655			

Average Rb = 3.53

TABLE : 4.6.2
BASIN MORPHOMETRY OF KOHIRA

Stream Order	No. of streams or segements	Bifurcation Ratio	T otal Length	Mean Length of segments in kms.	Cumulative mean length in km.	Length Ratio
u	Nu	Rb	kms	Lu	$\bar{Lu}$	RL
First	230		261	1.13	1.13	
Second	45	5.11	83	1.84	2.97	1.62
Third	12	3.95	47	3.91	6.88	2.12
Fourth	3	4.00	25	8.33	15.21	2.13
Fifth	1	3.00	13	13.00	28.21	1.56
		16.06	429			

Average Rb = 4.01

TABLE : 4.6.3
BASIN MORPHOMETRY OF LORSI

Stream Order	No. of streams or segements	Bifurcation Ratio	T otal Length	Mean Length of segments in kms.	Cumulative mean length in km.	Length Ratio
u	Nu	Rb	kms.	Lu	Lu	RL
First	90	4.09	115	1.2	1.2	
Second	22	4.40	35	1.6	2.8	1.33
Third	5	2.50	15	3.0	5.8	1.87
Fourth	2	2.0	12	6.0	11.8	2.00
Fifth	1		9	9.0	20.8	1.5
		12.99	186			

Average Rb = 3.24

TABLE : 4.6.4

BASIN MORPHOMETRY OF AUSANE

Stream Order	No. of streams or segements	Bifurcation Ratio	T otal Length	Mean Length of segments in kms.	Cumulative mean length in km.	Length Ratio
u	Nu	Rb	kms.	Lu	$\overline{Lu}$	RL
First	545		630	1.16	1.16	
Second	116	4.70	195	1.68	2.84	1.45
Third	27	4.30	112	4.15	6.99	2.47
Fourth	6	4.50	52	8.67	15.66	2.09
Fifth	1	6.00	28	28.00	43.66	3.23
		19.50	1017			

Average Rb = 4.88

TABLE : 4.6.5
BASIN MORPHOMETRY OF KOEL

Stream Order	No. of streams or segements	Bifurcation Ratio	T otal Length	Mean Length of segments in kms.	Cumulative mean length in km.	Length Ratio
u	Nu	Rb	kms.	Lu	$\overline{Lu}$	RL
First	189		187	0.99	0.99	
Second	44	4.30	76	1.73	2.72	1.75
Third	12	3.67	45	3.75	6.47	2.17
Fourth	2	6.00	12	6.00	12.47	1.60
Fifth	1	2.00	15	15.00	27.47	2.50
		15.79	335			

Average Rb = 3.99

TABLE : 4.6.6
BASIN MORPHOMETRY OF DURGAWATI

Stream Order	No. of streams or segements	Bifurcation Ratio	T otal Length	Mean Length of segments in kms.	Cumulative mean length in km.	Length Ratio
u	Nu	Rb	kms.	Lu	$\overline{Lu}$	RL
First	1261		1326	1.05	1.05	
Second	267	4.59	425	1.59	2.64	1.51
Third	55	4.85	190	3.45	6.09	2.17
Fourth	11	5.00	125	11.36	17.45	3.29
Fifth	5	5.50	35	7.00	24.45	0.62
Sixth	2	2.50	31	15.50	39.45	2.21
		22.44	2132			

Average Rb = 4.49

TABLE : 4.6.7
BASIN MORPHOMETRY OF KARSOTHA

Stream Order	No. of streams or segements	Bifurcation Ratio	T otal Length	Mean Length of segments in kms.	Cumulative mean length in km.	Length Ratio
u	Nu	Rb	kms.	Lu	$\overline{Lu}$	RL
First	242		294	1.21	1.21	
Second	51	4.75	80	1.57	2.78	1.29
Third	11	4.64	39	3.55	6.33	2.26
Fourth	2	5.50	41	20.50	26.83	5.77
Fifth	1	2.00	7	7.00	33.83	0.34
		16.89	461			

Average Rb = 4.22

TABLE : 4.6.8
BASIN MORPHOMETRY OF DHOBA

Stream Order	No. of streams or segements	Bifurcation Ratio	T otal Length	Mean Length of segments in kms.	Cumulative mean length in km.	Length Ratio
u	Nu	Rb	kms.	Lu	$\overline{Lu}$	RL
First	206		222	1.08	1.08	
Second	41	5.02	42	1.02	2.10	0.94
Third	10	4.10	29	2.90	5.00	2.84
Fourth	1	10.00	43	43.00	48.00	14.83
		19.12	336			

Average Rb = 6.32

TABLE : 4.6.9
BASIN MORPHOMETRY OF SURA I

Stream Order	No. of streams or segements	Bifurcation Ratio	Total Length	Mean Length of segments in kms.	Cumulative mean length in km.	Length Ratio
u	Nu	— Rb	kms.	Lu	$\overline{Lu}$	RL
First	356		419	1.2	1.2	
Second	82	4.3	132	1.6	2.8	1.33
Third	15	5.5	52	3.5	6.3	2.19
Fourth	2	7.5	61	30.5	36.8	8.71
Fifth	1	2.0	19	9.0	45.8	0.29
		19.3	683			

Average Rb = 4.8

TABLE : 4.6.10
BASIN MORPHOMETRY OF SURA II

Stream Order	No. of streams or segements	Bifurcation Ratio	Total Length	Mean Length of segments in kms.	Cumulative mean length in km.	Length Ratio
u	Nu	Rb	kms.	Lu	$\overline{Lu}$	RL
First	308		307	1.00	1.00	
Second	79	3.90	108	1.37	2.37	1.37
Third	14	5.64	67	4.79	7.16	3.49
Fourth	1	14.00	42	42.00	49.16	8.77
		23.54	524			

Average Rb = 7.84

b) The bifurcation ratio between one stream order to another is very high in case of some of the basins specially in the Sura II in which this bifurcation ratio between third and fourth order stream is 14. In the Ausane basin the *Rb* between the fourth and fifth order is 6, in the Dhoba basin it is 10 between third and fourth, in Koel it is 6 between third and fourth, in Sura I 7.5 between third and fourth and in Durgawati basin it is 5.60 between fourth and fifth order.

c) The study of Logarithms reveals the relationships of stream orders (u) with numbers (Nu), Lengths (Σ Lu) and cumulative Lengths ($\bar{L}$u) of streams of the Rohtas Plateau. In no case, the points form perfectly straight lines, i.e. the points show larger departures from the straight lines while a straight line is a good description of the sequence of points. So the validity of Horton's law of stream length (Strahler, 1971, p. 486) cannot be regarded as strongly supported in this case.

d) All these departures may have the following explanations :

i) The structure of the upper stratum of the Rohtas Plateau is homogeneous throughout from where almost all the first and second order streams emerge. Therefore, the *Rb* for the first and second order streams is between 3 to 5.

ii) In case of Goptha, Kohira and Lorsi basins where the *Rb* is within 3 to 5 for all stream orders, a perfect homogeneity of structure is illustrated.

iii) The Rb between 3rd and 4th order streams and 4th and 5th order stream is above 5 which shows that the streams enter into a different type of terrains on the lower reaches from a homogeneous terrain of the upper reaches of their courses. 'Abnormally high bifurcation ratios might be expected in regions of steeply dipping rock strata where narrow strike valleys are confined between hogback ridges' (Strahler, 1964, pp. 4-45). Accordingly the lower parts of the rivers are flowing in elongated subsided valleys produced by collapse of upper sandstone strata (Ch. 5.2). It has already been stated that almost all the rivers form waterfalls with sharp breaks of gradient and the *Rb* below the falls in almost all cases in high reaching to the highest figure of 14 in the elongated basin of Sura II.

iv) The study of stream ordering also conclues that the drainage basins which have normal range of *Rb* are of common shape, i.e., like pear or apple. The Kohira, Goptha, Lorsi, Ausane and Durgawati river basins (Fig. 4.6) are the striking examples.

(v) The drainage basins which have abnormal range of *Rb* in spite of the homogeneous structure of the surface are mostly long, narrow and elongated. This is due to the limited branching of streams. It happens when a large trunk stream is met with small unbranched numerous streams. The Sura I, Sura II, Karsotha, Koel and Dhoba river basins are such examples.

(vi) As regards size, the basins of normal *Rb* are generally larger in size than those which have abnormal *Rb*, provided that they are of the same hierarchy, i.e., either of fourth order or fifth order or so on. For example the Ausane basin and the Karsotha and Sura II basins are fourth order basins but Ausane basin is much larger than the later two basins.

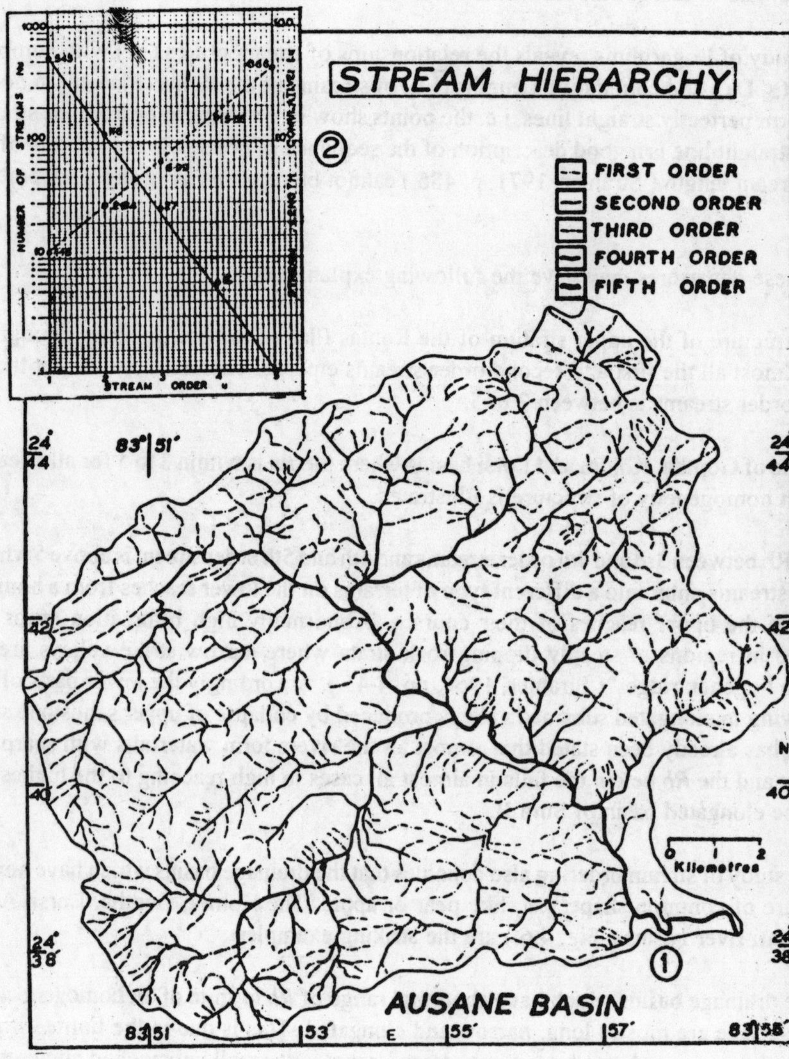

STREAM HIERARCHY
FIRST ORDER
SECOND ORDER
THIRD ORDER
FOURTH ORDER
FIFTH ORDER
NUMBER OF STREAMS
STREAM ORDER
2
1
AUSANE BASIN
Kilometres
0 1 2
N
24° 44'
24° 42'
24° 40'
24° 38'
83° 51'
53'
55'
57'
83° 58'
E

FIG. 4.6

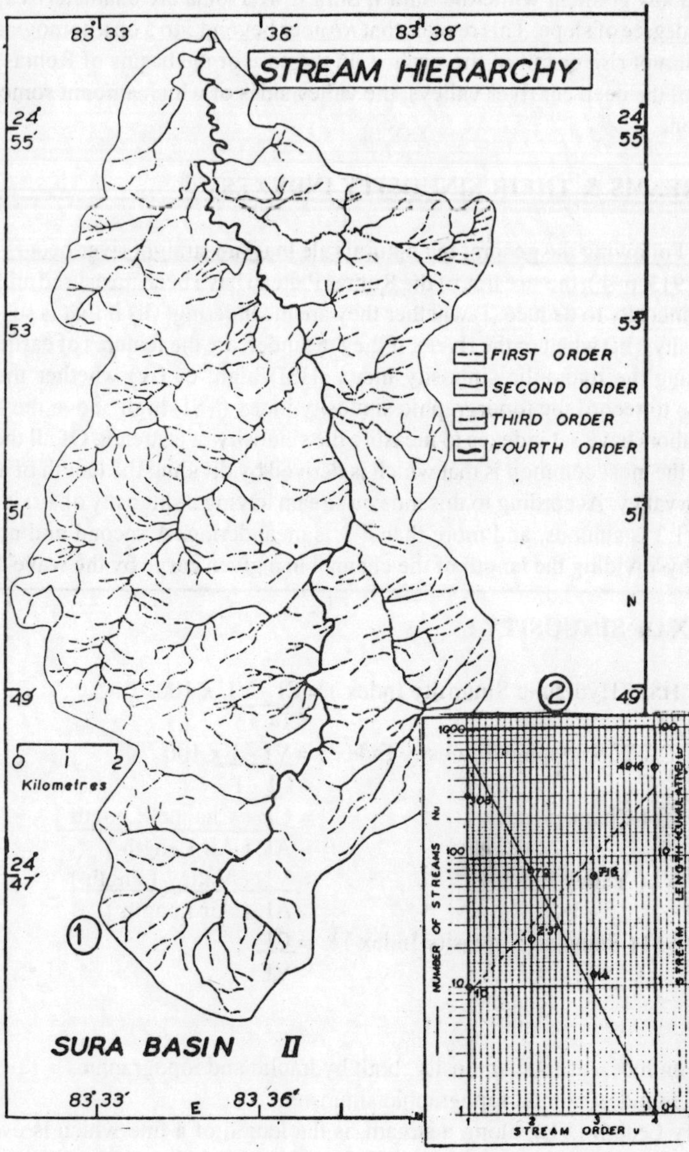
STREAM HIERARCHY
83°33'
36'
83°38'
24°55'
53'
51'
49'
24°47'
FIRST ORDER
SECOND ORDER
THIRD ORDER
FOURTH ORDER
N
0 1 2
Kilometres
①
②
SURA BASIN II
83°33'
E
83°36'
NUMBER OF STREAMS
STREAM LENGTH (CUMULATIVE)
STREAM ORDER u

FIG. 4.7

(vii) Morphologically also the basins of normal *Rb* are different from those of abnormal Rb. The Ausane, Koel, Goptha, Kohira, Lorsi and Durgawati basins are comparatively of low relative relief and low gradient while the Sura I, Sura II, Karsotha are characterises high relative relief and high degree of slope. This reveals that *Rb* goes beyond 3 to 5 on a homogenous structure when there is abrupt rise or fall of the surface. In the case of the basins of Rohtas Plateau it happens because of the deep cut river valleys, the valley sides of which amount sometimes to more than 90^0 of slope.

4.6. STREAMS & THEIR SINUOSITY INDEXES

Following the general but natural rule that ' no straight river occurs in a state of nature' (Dabis, 1913, p. 4), the streams of the Rohtas Plateau have been measured and analysed in regard to their sinuosity to deduce (i) whether they are meandering; (ii) if that is so, what is the amount of tortuosity; (iii) whether the rivers, if they meander, are the features of earlier planation surface maintaining the hydraulic sinuosity index (HSI) high; or (iv) whether they have developed otherwise to record the toporgraphic sinuosity index (TSI) high above the HSI.It is noted that many authors have set indexes to measure the sinuosity of a stream. Of all the existing sinuosity indexes, the most common is that which is derived by dividing the length of a reach as measured along the valley. According to this index a stream having a sinuosity of 1.0 is straight, just above unity to 1.3 is sinuous, and more than 1.3 is meandering. A second and more recent index is derived by 'dividing the length of the channel in a given curve by the wave-length of the curve'

*** INDEX OF SINUOSITY :**

HSI (Hydraulic Sinuosity Index) = $\frac{CI - VI}{CI - 1} \times 100$

TSI Topographic Sinuosity Index) = $\frac{VI - 1}{CI - 1} \times 100$

CI (Channel Index) = $\frac{\text{CL (Channel Length)}}{\text{AL (Air Length)}}$

VI (Valley Index) = $\frac{\text{VL (Valley Length)}}{\text{AL (Air Length)}}$

SSI (Standard Sinuosity Index) = $\frac{CI}{VI}$

N.B. :

1) CI is an index of total sinuosity, both hydraulic and topographic.
2) VI is an index of total topographic sinuosity.
3) Valley Length (VL), along a stream, is the length of a line which is every where midway between the base of valley-walls. It will equal Channle Lenght (CL) whereever the valley walls descend, directly to the water's edge, and will be less than CL wherever a flood plain has developed.
4) Air Length (AL) is the shortest distrance between two points of a meandering bend of river (for big rivers) and a shortest distance between the source and mouth of the streams in case of the small rivers.

TABLE : 4.7

SINUOSITY INDEXES (MUELLER'S MODEL)

Name of River	VALLEY LENGTH (VL)			Channel Length (CL) kms.	Air Length (AL) kms.	C.I.	V.I.	H.S.I.	T.S.I.	S.S.I	Gradient in metre/ kms.
	Left Bank kms.	Right Bank kms.	Average Length kms.								
Ausane	53	51	52.0	51.0	24.0	2.13	2.17	-4.00	104.00	0.98	8.6
Dhoba	52	52	52.0	52.0	38.0	1.37	1.37	0.00	100.00	1.00	7.8
Koel	40	39	39.5	40.0	33.0	1.21	1.20	5.00	95.00	1.01	8.0
Sura I	80	79	79.5	80.0	42.0	1.90	1.89	1.00	99.00	1.01	4.7
Sura II	53	53	53.0	53.0	37.0	1.43	1.43	0.00	100.00	1.00	6.5
Goptha	39	38	38.5	38.0	31.0	1.24	1.24	-4.00	104.00	0.99	8.5
Karsotha	47	47	47.0	47.0	36.0	1.31	1.31	0.00	100.00	1.00	7.6
Durgawati	28	29	28.5	28.5	22.5	1.27	1.27	0.00	100.00	1.00	16.8
Lorsi	23	24	23.5	23.0	12.0	1.91	1.95	-4.40	104.40	0.98	15.9
Kohira	20	21	20.5	20.0	12.0	1.67	1.71	-5.00	105.00	0.98	15.5

N.B. :

C.I. = Channel Index = CL / AL

V.I. = Valley Index = VL / AL

H.S.I. = Hydraulic Sinuosity Index = $\frac{CI - VI}{CI - 1} \times 100$

T.S.I. = Topography Sinuosity Index = $\frac{VI - 1}{CI - 1} \times 100$

S.S.I. = Standard Sinuosity Index = $\frac{CI}{VI}$

(Leopold, et. al., 1966, p.60). An index of sinuosity is ' derived by dividing the thawleg, or the line of maximum depth, by the length of the valley' (Leopold, et. al., 1960, p. 133). These indices do not consider the role of surface configuration or topography in directing the course of the stream. Noting some inherent flaws in these existing sinuosity indices, Mueller (1968, pp. 371-385) has derived a new sinuosity index recently and in that he has given due importance to the channel and surface configurations in making the sinuous bends of the rivers. Mueller's insistence on including the role of the topographic factors in the computation of sinuosity indexes is remarkable, as it gives a new dimension to the concept of sinuosity. According to him, the turtuosity of the course of a meandering stream is the outcome of both topographic and hydraulic factors and is expressed by a ratio called the index of sinuosity.*

In the Rohtas Plateau, measurements have been taken in the 10 river basins; like Durgawati, Ausane, Karsotha, Sura I, Sura II, Dhoba, Koel, Goptha, Kohira and Lorsi. Both the hydraulic sinuosity index (HSI) and topographic sinuosity index (TSI) of these important rivers have been determined from the 1:50000 Topographical Sheets. The data has been tabulated in Table 4.7. Some long and cross profiles have also been drawn for the purpose.

The study reveals the following points :

1) The data (Table : 4.7) reveal that the reaches of the streams Koel, Goptha, Durgawati and Karsotha in the Plateau are not meandering. They are just sinuous.

2) Other streams like Ausane, Dhoba, Sura I, Sura II, Lorsi and Kohira appear to have meandering courses on the basis of data but in most case CI and VI are more or less the same concluding thereby that the streams have not developed flood plains in spite of their turtuosity.

3) Since these streams are bounded on both sides by steep rocky walls and their channels are very much constricted the TSI (Topographic Sinuosity Index) is very high above the HSI (Hydraulic Sinuosity Index), sometimes 100 per cent and in some cases more than 100 per cent. This suggests that the turtuosity is not of late mature or old stage, but it is exclusively of youth stage. The profiles drawn longitudinally and transverse to the streams reveal the same fact because they show greater steepness in gradient.

4) The hydraulic sinuosity index (HSI) is not high in most of the river reaches. The HSI ranges from -4 to 5 for all the measured rivers of the Plateau. This shows that the rivers are not flowing on flood plains.

5) The data (Table : 4.7) show some interesting character of streams particularly in case of Ausane, Gophta, Lorsi and Kohira which record higher VI than CI. In most cases of meandering streams, channel length or index is greater than the valley length or index, and consequently SSI becomes more than 1. Greater the SSI the intensity of meandering is higher. But in exceptional cases VI is recorded greater than CI, and it happens only when a deep cut meandering stream is bounded of indented valley sides. Such cases emerge either when main meandering stream is joined by numerous deep cut tributaries or when valley sides are subject to collapse due to weak stratum below. In case of the streams of the Rohtas Plateau most of the valleys particularly Ausane, Goptha, Lorsi and Kohira have deep-cut meandering beds due to collapse of materials. This suggests that a considerable number of caverns was developed during the geological past which collapsed during the Tertiary movement. This also suggests that these meanders are not incised or entrenched over the earlier flood plain, but their development is quite different from rest of the Plateau rivers of Chotanagpur.

5

SURFACE FEATURES & KARST LANDSCAPE

The main objective of this book has been to bring unexplored features of the Rohtas Plateau into light and make a discussion of their genesis geomorphologically. The Rohtas Plateau is not only presenting a high land bounded by scarps, but it had numerous fascinating landforms, each speaking their own characteristics and origin. If we take a single feature, say, scarp, it is sufficient to attract attention of the scientists of geology, geomorphology and other allied desciplines as well as the common observers. The highlands of Chotanagpur have a considerable length of escarpments on the margins of a series of Plateau, but they are not so steep and inaccessible as is the escarpment of the Rohtas Plateau. Similarly deep-cut valleys with almost vertical rocky walls on both the sides are confined only to the Rohtas Plateau in the whole of the Bihar Province. Other features like **Khohs** and notches, waterfalls, caverns and karst features therin and hills and hillocks are really uncomparable to many features of the adjoining highlands of Chotanagpur. This particular chapter has, therefore, been devoted to discuss and analyse all these geomorphic features of great interest.

5.1 THE FLOOD PLAIN LYING BELOW 140 METRES

The river Son which flows at the southern margin upto the longitude 83^0 55' east takes a north-easterly course from this longitude. This river has produced flood plain on both the sides of which the left side is within the delimited area of study.

The western segment of the flood plain stretching from the western margin to the longitude 83^0 50' east is very narrow while it goes on increasing in width north east-wards (Fig.5.1). Its width varies from a half kilometre to one kilometre on the south while in the eastern segment it widens up to 3 to 8 kilometres. There are encroachments of the Plateau at places towards the Son particularly in the western parts which make the Flood Plain narrow; but it widens at those places where streams come down from the Plateau to join the Son. Virtually these are the places where the width of the flood plain is greatest ranging from 5 to 8 kilometres. The area is below 140 metres in height and in fact the left margin of the flood plain towards the edge of the Plateau has been determined by 140 metre contour.

The surface of the flood plain is almost having a gradient of below 2 (Fig. 7.3). There are, however, some breaks in monotony, where isolated small hills rise above the level of the flood plains. These hills are the Bhawnathpur, Amarkha and Murali having a height of 150 m., 154 m. and 286 m. respectively. Since they rise above the flood plain the absolute height of these hillocks above the level of flood plain is 10 m., 14 m. and 146 m. respectively. The flood plain

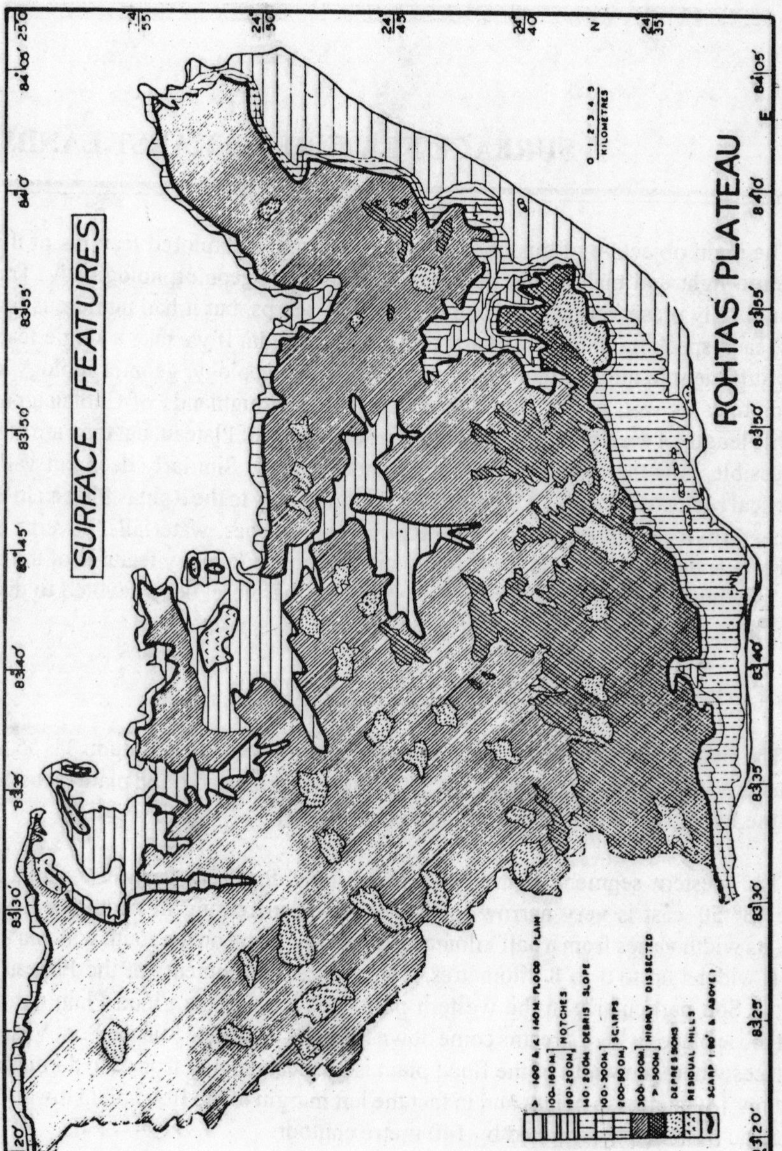
SURFACE FEATURES
ROHTAS PLATEAU
KILOMETRES
FLOOD PLAIN
NOTCHES
DEBRIS SLOPE
SCARPS
HIGHLY DISSECTED
DEPRESSIONS
RESIDUAL HILLS

FIG. 5.1

has very low gradient towards the Son as well as a gradual observable sloping from west to east following the course of the Son river. This unit of the study area covers, an area of 147 km^2 which represents 6.28 per cent of the total area.

The flood plain seems to have originated owing to the combined processes of weathering, mass wasting and fluvial action., The scarps originally formed due to faulting in the Tertiary period are receding backwards because of the processes of weathering, mass wasting and other denudational activities. The talus which was transported from the scarp face and the Plateau top forms the looser foundation of the flood plain where rock fragments of varying sizes are found buried under a thick mantle of river-borne materials. The coarse materials received from the Plateau and the fine materials transported with the son water particularly during the floods produce alternating layers of coarse and fine materials in the flood plain. Again this combined process makes the ground almost level. Consequently, the plain looks fairly level close to the river with gradual increase of slope towards the Plateau.

5.2 THE NOTCHES, KHOHS AND LOW LYING RIVER VALLEYS

A conspicuous feature in the Plateau of Rohtas which catches eyes of even a casual observer is the long elongated deep river-valley having finger-like projection towards the source region. This feature, developed mostly along the northern and south-eastern edge of the Plateau is very akin to gorges and ravines. It is locally popular as **khoh** because of its peculiar shape. The term notch appears to have come owing to its finger-like projection. These ***khohs*** and notches are occupying the major portions of the lowlying river-valleys which vary in height from 100 m. to 200 m. in the south and south-east.

The notches, ***khohs*** and low lying river valleys are the result of headward erosion by streams descending down from the Plateau surface. 'Sizeable river like Durgawati, Sura, Karamnasha and Karsotha along with their tributaries have cut deep notches on the scarp while descending down on the plains in the north so that the upper courses of each of the streams have formed scarp-bound valleys of various shapes and indentations' (Dutt, 1971, p.231). On the basis of a close observations by the author there is, however, a slight difference between notches and ***khohs***. Notches are those indentations which extends like finger into the body of the Plateau where streams emerge from spring. The river erosion is regarded to be responsible for the origin and development of notches. ***Khohs***, on the other hand, have a more complex origin. The area is composed of a succession of sandstone and limestone where sandstone overlies the limestone. The water filters down the sandstone without initiating any chemical changes. But by the time the water reaches the limestone and shale layers chemical proceses start and solution of limestone also begins. The water reacts on calcium carbonate and dissloves it and slowly the solution drains out through tiny caves developed on the sides of the scarp face. Gradually the process of sloution leads to inlargement of caves and removal of the entire body of limestone underlying the sandstone bed. Devoid of any support from below the overlying sandstone structure subsides into the caves and cave becomes open such a feature is locally known as ***khoh***. The **Amba khoh, Gularia khoh, Rakim kund khoh etc**.are some of the striking examples.

The prominent characteristics of these **khohs** are related to the gradient of the long profile particularly at the source region. Generally near the source, the long profile of a river has a low gradient, but in cases of all these streams the gradient at the source region is very much steep and abrupt. One interesting thing is that the Lal Darwaja of Rohtas Fort (Western Fort Gate) and other fortwalls are assumed to be constructed round about 1000 years ago. If the history shows that it is only of Mughal period , even then the eroded portion might have receded and it can be said that the water divide area between the **Gularia khoh** and its opposite side might have been narrower and narrower. But after the lapse of a long period, there is no sign of such recession. On the basis of the nature of long profile of these streams, steep and bare rocky valley sides and slow rate of erosion, it may be said thet the *khohs* might have been karst valleys which developed due to the formation of caverns and ultimately due to submergence and removal of all the materials.

Thus all the notches are nothing but the karst valleys. Now a question arises why all the areas are not submerged. The answer may be that the quality of limestone differs from one region to another and also the amount of water and nature of porosity differs considerably. So, caverns would not have developed in all the parts beneath the sandstone capping. Wherever, these would have been developed they are at present existing in the form of **khohs.**

The low lying river valleys are those which are at lower height in comparison to the height of the Plateau. These are the end product of headward extension of erosion. Therefore, such features can be classified into two groups depending upon the surface elevation.

5.2.1. 100 M. AND 160 METRES

In the northern portion of the Plateau, the area covered by notches, *khohs* and low lying river valleys is greater in comparison to those of the southern edges. This is because many major rivers originate on the southern margin of Plateau and flow towards the north cutting across the scarp face. These rivers with **khohs** & notches are Kohira on the extreme west and Sura, Kanjiari, Karsotha, Nyuri, Koel, Dekura, Durgawati, Goptha, Chormara, Dumarkhar and Dhoba on the east respectively. The height of the river valley varies between 100 metres to 160 metres. The fingerlike projections some of which end up in **khohs** are higher in elevation averaging 140 to 160 metres. Since the general level of the Plateau has an inclination from south to north, all the major streams flow towards the north. River Durgawati has its origin close to the southern scarps but river flows away towards the north and away from the river Son. Almost all the rivers have formed waterfalls in the **khoh** regions where surface run-off is supplemented by seepage of groundwater on **khoh** side.

5.2.2 160 METRES TO 200 METRES

The area of the scarp facing river Son in the south is highest of all the Plateau regions and spot heights show elevations up to 641 metres. The height gradually decreases towards east and northeast. Therefore, the scarp face runs like an unbroken wall almost parallel to the course of the Son river but at slightly varying distances from it. Since the Plateau slopes towards the north, no major streams could develop discharging their water into the Son. Only small rills or **nalas** which originate at the scarp drain towards Son crossing debris slope. But two rivers are of significant dimensions and deserves special considerations because they are the only rivers flowing towards the south and discharging water into the Son. They have also produced notches and **khohs** and associated valleys in the southern part of the Plateau. The first of the river dissected area lies along longitude 83^0 40' east where Bhawatia river descends down the Plateau surface. This is a minor streams, very small in length but it has produced a **khoh** in the otherwise unbroken wall of the southern scarps. This **khoh** is called **Bansi khoh**. The height of this **khoh** and the notches extending towards Plateau ranges between 160 m. to 200 metres.

The second one, somewhat larger in area of this feature is found in the valley of river Ausane. This river has carved notches and **khohs** which extend deep towards the south, west and north. Such **khohs** are called by various names like **Amba khoh** in south-west and **Rakim khoh** towards west. The area between 160 to 200 metres high which forms a part of this group is found at a distance of about 1 kilometre from the foot of the scarps in the northern valley of the river and at a distance of one-fourth kilometre in the western part. The distance of western notch is about 1 kilometre, which suggests a gradual increase of height from 200 metres to 500 metres, the height of the scarp-face being 500 metres. The area has a shape of inverted y.

5.3. THE DEBRIS SLOPE RISING BETWEEN 140 M. AND 200 M.

The debris slope along the southern margin of the Plateau is wider and very conspicuous and it extends between the river Son and the Plateau scarp. In the north-eastern and northern side the debris between 140 m. to 200 metres extends as a narrow ribbon following the scarps with complete fidelity. 'It is noticed that the loose material produced by weathering and erosion in the Vindhyan Plateau especially the Plateau composed of shales with a thick capping of upper Kaimur sandstones move downward and settles in the lower part of the Plateau slope' (Rai, 1971-71, p.172). The origin of the debris slope can be attributed to the downward movement of weathered material above the plateau surface. The upper portion is composed of sandstone while the lower portion is made up of shales. Shales break-up easily under the impact of rainfall and associated decomposition and desintigration. The loose material creep down under the force of gravity and accumulate along the border regions of scarps. Where streams descends down from the Plateau surface these materials are carried away by moving water. But where streams are absent the weathered material goes on accumulating and producing a wider debris slope area. Such is the case along the bank of the river Son. The process has been illustrated to some extent, on the bank of the Son the soil is a firmer tilth of fine sand and clay where the proportion of clay decidedly increases downstream from any confluence with the left bank tributaries. Obviously the supply of this additional clay is a function of the degradation of the shales of the Rohtas Stage by these left bank streams. This sandy-loam bank, susceptible to regular inundations from the Son during seasons of floods, extends like a terrace of persistent width and ends against a still higher land surface formed of boulder and pebbles of various sizes and descriptions. The boundary is essentially parallel to the channel of the Son. Although the relative relief of the sandy-loam terrace and the river-bed suggests some order of incision, thses two landforms are mutually dependent functions of alluviation of the Son (Chakraborty, 1965, pp. 109-110). Shale becomes slippery when it gets soaked in water and allows rapid transportation of weathered materials produced by process of mass wasting. This process of downward movement can also be described as gravity sliding which is instrumental in the movement of materials from Plateau surface to

TABLE : 5.1

NOTCHES & KHOHS

Name & Location	Lat. & Long.	Length	Width in Metre			Height in Metres			Straight wall/indented
			Source	Middle	Lower	Source	Middle	Lower	
Sura I Jawarh Khoh	24°50' - 24°55' 83° - 83°30'	5 km.	50	700	1000	180	125	120	Straight but indented at some places
Sura II Ropar Khoh	24°49' - 24°53' 83°38' - 83°36'	4 km.	30	500	1500	170	140	100	Very indented some small notches are 1 km. long
Karsotha River	24°46' - 24°49' 83°37' - 83°39'	3 km.	75	500	1500	220	200	140	Straight
Koel River Bhurli Khoh	24°47' - 24°48' 83°42' - 83°43'	2 km.	40	250	500	180	120	105	Straight
Kalanwa River	24°40' - 24°42' 83°46' - 83°45'	2 km.	50	70	150	260	230	180	Straight
Amba Khoh	24°38' - 24°39' 83°55' - 83°56'	1.5 km.	50	1000	1200	1200	230	190	Straight V. shaped

the foot of scarps.

The widht of debris slope in the western portion is about 1 kilometres and lies close to Son river (Fig. 5.1). The height of this slope is below 200 metres. It is succeeded northward by a slope of more than 200 metres high which is same in character but slightly different in composition. In the lower part **i.e.** between 140 m. to 200 m. the slope is composed of somewhat finer materials but the slope just close to the scarps having more than 200 metres high consists of large boulders and pebbles. This gives an impression of terrace like slope rising gradually from the bank of the Son.

East of longitude 83° 38' E. the debris slope goes on widening. The slope is gradual and very gentle. The width varies between 3 kilometres to 7 kilometres. The monotony of the slope is broken by a few isolated hills which rise between 83° 45' and 83° 50' east. These hillocks having west-east elongation are completely detached from the main Plateau. They are Tiara, Parchha, Katudanr, Tenwa, Banua and Dumaria having heights of 188, 198, 192, 165, 185, 170 metres respectively.

This slope becomes very narrow at the south of Ausane river where flood plain of the Son has extends deep into the valley of its tributary. It extends inwards like a tongue up to a distance of about 5 kilometres from the mouth of the river.

In the north easrern part the width of the narrow debris slope is no where more than 1 kilometre. It is about a half kilometre wide in the north eastern portion along the bank of the Son; because here the flood plain has the greatest width and river-borne deposits cover the layer of debris slope.

The northern scarps are irregular in extent and, therefore, the ribbon of debris slope is irregular, which follows the foot of the scarps.

5.4 THE SCARP FACE

Of all the features in the Plateau of Rohtas, scarps (Plate : 2) are the most conspicuous and eyecatching. The highlands of Chotanager have a considerable alignments of scarps that separate two plateau masses of different elevations, but those scarps, being more precipitous and granite gneiss shield, are very much modified. So, the steepness has gone from many areas. The case is different in the Rohtas Plateau. The scarps here are very steep, sometimes more than 90° (Plate 5.2), which project Vindhyan sandsotnes outward on the top. The limestone and shale beds below are eroded quicker than the overlying sandstones. It is because of this that the scarps of the Rohtas Plateau are very steep forming a fort-like walls for a long distance. In many areas they cannot be tamed.

The Plateau has escarpments on three sides (Fig. 5.1) on the north, south and east. It is noted that the Plateau is structurally such that the escarpements will amount 90° and more of gradient, and consequently such a gradient is maintained in most parts of the alignment. It is why the Plateua top cannot be linked with communication lines through these portions. The modified

PLATE : 2. NATURE OF SCARPS

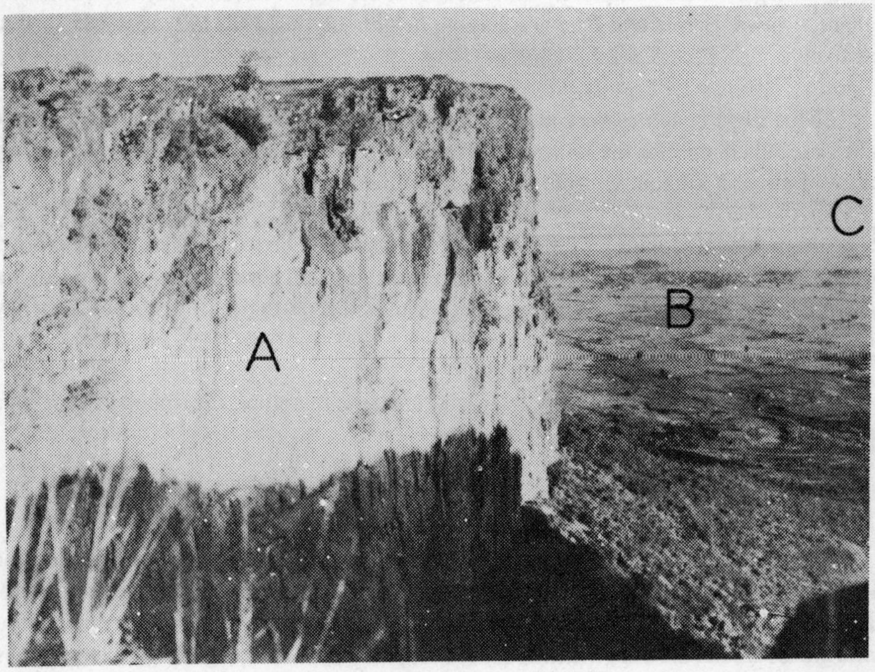

Bold Scraps (A) at almost 90^0 rising above slightly sloping alluvial tract (B) on the bank of the Son river (C) east of Baulia

Plate - 2

portions of the scarps are observed only in those areas where there were caverns in the geological past and those caverns collapsed following the earth-movements resulting **khohs** discussed earlier. Due to collapse of caverns the materials of ceiling came down on the floor making bumps close to the steep-sided rocky walls of the karst valleys. It is through these bumps or debris slope lying in the **khohs** that one can ride on the Plateau. Since thses resistant to erosion for a long time. It is, however, noted that such areas are limited around the Plateau and, therefore, **ghats** to ride on the Plateau are few in number (Fig. 5.1). That the valleys or **khohs** have sandstone bumps has been proved by numerous soil profiles taken in the valleys.

According to the elevation the scarps have been categorised into two, the northern and the south-southeastern scarps.

5.4.1 THE NORTHERN SCARPS OF 100 M. - 300 M.

The scarp faces on the north of the Plateau are comparatively low in height and deeply indented. Most of the **khohs** intrudes through thses scarps. 'Just below the town Sasaram the scarps takes a turn westward and follows a sinuous course for about 48 km. to form the northern edge. The northern edge of the Plateau is very different from the southern edge' (Dutt, 1971, p.231). ' As a result of the northerly predominance of the drainage, the northern scarp of the Kaimur Plateau is deeply indented' (Pascoe, 1959, p.532). The alignment of the northern scarps has been very much disturbed through the geological periods. It can be inferred that the scarps might have been straight having west-east elongation and overlooking the lower portions of the north. Since the Plateau got tilted towards north, the drainage sysstem, both surface and underground, was dominant towards north characterized with a bunch of tributaries, large volume of water nad high velocity of streams etc. The underground water developed karst features and the surface drainage accelerated the pace of erosional activities. Consequently, the northern scarps receded and became indented owing to the collapse of caverms subsequently.

5.4.2. THE SOUTH-SOUTH EASTERN SCARPS OF 200 M. - 500 M.

The southern slopes are precipitous and form a series of escarpments to which the name Vindhyan is more appropriately applied. The scarps of the southern part facing the son are more regular and straight because rivers have not broken its continuity from west to east. 'The scarp face is extraordinarily continuous and runs in a direction sympathetic, to the coarse of the river Son' (Dutt, 1971, p.231).

It is because of the fact that 'no stream of any significance except perhaps the river Ausane has broken through this wall and joined the river Son as its tributary' (Dutt 1971, p.231). Minor indentations in the line of scarps have been produced by Bhawatia river in the west and Ausane river in the east. The break is narrow along the former while it is wider along the latter. The scarps continue as a smooth line north-east of 83^0 57' E.

CHARACTERISTICS & ORIGIN OF SCARPS

The scarps on north of the Plateau are comparatively low in elevaticn but they are also prominently straight and vertical having sharp junctions with the level cultivated areas adjoining the Plateau. Viewed from the Grand Trunk Road which runs north of the Plateau, the scarps seem to extend for a number of kilonmetres parallel to the road. At places where the rivers debouch from the Plateau steep ravines have originated cutting through the escarpments and bringing the Plateau area closer to the level of adjoining lowlands. The areas away from river courses are still maintaining a bold and vertical scarp face. The stratigraphic sequence of the sedimentary strata is clearly exposed on these scarp faces which tell the geological structure of the Plateau like a book.

The scarps run for 78 kilometres from Kosdihra to Baulia on the south; 43 km. from Baulia to Tarachandi on the east and 146 km. from Tarachandi to Ramgarh on the north. The air length of these segments is 38, 40 and 74 kilometres respectively. This reveals that the southern scarps are comparatively less arcuate and the eastern is more and the northern one has the maximum inundation. This is because most of the streams originate from the southern margin and flow down to the east and north following the general gradient towards thses two directions. We have noted that the Plateau has been subject to tilting. (Ch. 3.1.2).

Structurally the scarps are heterogeneous and layers of different formations peep through the scarps. The general succession is from Kaimur sands one on the top to limesotne and shale on the bottom (Ch. 2.2 & Table : 2.1). Since the Plateau is tilted the beds of sandstone, limestone and shale are apparently observed on southern and eastern faces of the scarps while on the north only sandstone bed is lying above the ground at many places and the lower beds are projected towards the interior below the general level of the ground. The limesotne and shale beds are, however, observed where the streams have cut the surface deep. Such stances are common along the notches and **khohs** throughout the Plateau.

These scarps were originated primarily due to faulting in the Archaean period. Since then it was never planated up to the sea level. The subsequent movements resurrected the falulting and uplift. This is why it is falult scarp. Fault scarps are those which are, 'a direct product of fault displacement' (Fairbridge, 1968, p.322). There has been, however, a case of recession of scarps from their original place towards the interior. The structure has been responsible for this recession. The lower beds of limestone are subject to solution and subsequently, upper thick and heavy beds of sandstone have collapsed allowing the denudational agents to wash out the materials. This recession will continue as long as the Plateau mass is there. At the same time the steepness of these scarps (90 and more) will be maintained because of this sturcture which will always facilitate the limestone to solution and upper sandstone to collapse. Thus it will never have strong base below and the strata above will remain the scarps in modified form.

Thus the scarps of the Rohtas Plateau will stand as fascinating landform as long as the Plateau is there. We can say in addition that there is more structural control on the origin of these scarps than any other processes.

5.5. THE HIGHLY DISSECTED PLATEAU

The Plateau top is a highly dissected surface which is completely uneven all around. It consists of mesas, buttes, cuestas, small hillocks and interspersed depressions (Fig. 7.1).

Therefore, the versions that 'the most outstanding feature of the Plateau is the remarkable flatness of the surface' (Dutt, 1971, p.231) or the 'Plateau top is generally flat' (Chakraborty, 1965, p.113) are totally baseless. It is noted that the area has never gone under the sea after its formation like the plateau masses of adjoining Chotanagpur highlands. But it remained high through the geological ages and was never planated to the base level as we find the case in the higlands of Chotanagpur. The geological records also show that the Upper Vindhyan formation **i.e.** the Kaimur sandstone which forms the Plateau top under discussion does not consist of any intruded rocks. Therefore, whatever undulations and hill or ridge-like features are there, they are not stems of previous formations nor they are monadnocks or inselbergs. The hills and ridges are but the left out portions by the streams and rivers. Thus we can say that differential denudational processes have been responsible for these uplands, hills, hillocks, and ridges of Vindhyan sandstone. As regards the depression like features the author concludes on the basis of his close observation and field study that they are initially the area of relatively higher percolation of water which went ultimately to the limestone bed below. It appears that some short of subsidence might have occured in these portions which facilitated the accumulation of water in the circular or oval shaped depressions first and then initiation of a drainage line towards the lower elevation. The notches and **khohs** which intrude into the interior of the Plateau are virtually the result of regular action of surface and underground drainage.

The Rohtas Plateau evolved in the Pre-Cambrian times. Since then it was uplifted, faulted, warped, tilted and was subject to compressions and tensions owing to the earth-movements particularly the Hercynian and Tertiary and other endogenetic forces. As a result the Plateau top under discussion was tilted towards north and east and joints of all scales developed in the plateau. This phenomenon has evolved the drainage lines mainly from south to north and through the joints. Therefore, the large river valleys developed along the joints leaving uplands, ridges and hills in between them. The structural control has developed depressions and caverns which later collapsed to form the steep-sides narrow notches and **khohs** within the Plateau. All these phenomena have made the Plateau top highly dissected. This character of the Plateau can be recorded only on the basis of intensive field study and after going through the new topographical maps. Therefore, the feeling of many earlier workers that the Plateau is flat is absolutely wrong.

The material of the Plateau top is the same having sandstone capping throughout the area but it differs in height from area to area. According to height it is divisible braoadly into two parts :

5.5.1 THE PLATEAU TOP BETWEEN 300 M. AND 500 M.

The Plateau top lying between 300 metres to 500 metres is comparatively larger and cover 1135 kms^2, 48.51 per cent of the total surface area of the Plateau. At places, rivers have intruded into the Plateau and carved out narrow nothces in its body. The height is still lower. But on the Plateau proper there is a gradual slope from 500 m. to about 300 metres which is the height of northern scarps.

This physiographic unit is vast in area lying roughly north of the latitude 24^0 40'. It has, however, made an extension towards the south in the form of drainage basins. This unit is highly dissected due to the middle and lower reaches of streams, inter-fluvial uplands, numerous depressions and extension of notches and **khohs** into the Plateau.

As noted earlier the undulations of this physiographic unit are marked by hill or ridge-like features, which are generally elongated from south-west to north-east, almost parallel to the primary and secondary joints. The unit is drained mainly by the Karamnasha, Sura, Gurwat, **Okhargara nadi, Jigna nala, Lorsi nadi, Kohira nadi** (Toposheet No. 63 P/5); Dekura *nadi*, **Koel nadi, Nyuri nala, Karsotha nadi** (63 P/9); Hasadah, Hathidah, Khora, Nachan, Badiya, Bajra (63 P/10); Amlo, Kachhuhar, Dumarkhar, Goptha, Chormara (63 P/13); Ausane, Barua, Gayghat, Gularia (63 P/14) and Dhoba and Tutla river (72 D/1). All these streams flow almost parallel to each other with a good net of tributaries. In between these streams lie the elongated uplands which appear to be ridges. Sometimes these uplands have radial pattern of drainage which facilitates denudation in all sides producing hill-like features. It must be remembered that the materials that make the upland features are the same as those of the surrounding areas. The old Vindhyan red sandstone have broken horizontally producing wide and long sheets lying over each other. Such sheets have been subject to weathering particularly mechanical one. As a result ridges and hills are composed of separated sheets and broken angular blocks. Naturally the upland features are not conical or pointed nor dome-shaped; but such features are sufficient to present a dissected view of the Plateau top.

5.5.2 THE AREA ABOVE 500 METRES

The area above 500 metres high lies in the southern half of the Plateau. Here the height is greatest. There are numerous depressions in this part also. Groundwater appears as a spring in these depressions which sometimes takes the forms of small streams. The 500 metres contour line which divides the Plateau runs very irregularly from west to east. The contours have receded south-wards where streams have extended on the Plateau surface by headward erosion. They have helped in reducing the height of the Plateau and are responsible for the irregular shape of 500 metres contour line.

As noted earlier (Ch. 5.1.5) in this phsiographic unit is marked hill or ridge-like features, which are also generally elongated from south-west to north-east. This unit is drained mainly by the Durgawati, Karamnasha, Hasadah, Hathidah, Khora, Nachan, Gurwat, Bajara, Badiya, Titahi, Bhawatia (toposheet no. 63 P/10) and **Barua nala, Gayghat nala, Amba khoh nala, Gularia khoh nala, Ausane nadi, Haghra nala, Sarsawati nala, Sanha nala, Boha nala** (63 P/14). All these streams flow almost parallel to each other. In between these streams lie the elongated uplands which appears to be ridges or hills. Important among the upland features are Sohdag (602 m./24^0 33' 30" : 83 41'), Dumurka (637 m./24^0 32' 30" : 33 34'), Khukhuman (596 m./24^0 34' : 83^0 44'), Gamaharia (571 m./24^0 38' : 83 39'), Dumarkhoha (641 m./24^0 33' : 83^0 32'), and Dhansa (551 m./24^0 40' : 83^0 40' 30").

5.6 DEPRESSIONS

As mentioned earlier (Ch. 5.1.5) the Plateau top of the study area comprises of a number of low lying area, generally saucer-shaped with one outlet in the form of stream or **nala** oozing out of spring. This is peculiar feature clearly observable on the topographical maps in the form of wide patches. Such low lying areas appear to be evolved due to percolation of water towards the interior up to limestone beds. Sometimes it is questionable whether percolation is possible into the sandstone capping of about 250 m. to 350 m. A close observation of such areas by the author compels to believe that such low lying oval and circular areas are nothing but the subsided portions. It is a fact that Plateau top has developed joints and cracks due to earth-movements through the geological period. Had there been a thin capping of säŋdstone over the limestone beds, stone bgeds, the entire area might have developed karst features right from sink-holes, dolines and uvalas to karst valleys. The thick capping could not make it possible but it could not deny the water percolation into the interiors. This could make possible a subsidence atleast due to the formation of caverns below the sandstone bed and subsequent earth-movements.

Such depressions comprise of water bodies and sub-surface drainage which oozes out in the form of springs. Consequently, these are the areas where most of the settlement sides can be observed. Only such areas have water availability, though in scant quantity. Whatever agricultural practices both cultivation and cattle-rearing are observed on the Plateau top, these are on such depressions. From these low lying areas uplands are viewed as ridges and hills.

The depressions are scattered throughout the Plateau. Some of the important areas are Dighar, Chaudharna, Semra, Kamharia, Karkatgarh, Jigni, Bhaluburan, Barap, Bahadag, Dahar, Kham Kalan, Salea, Barwan Khurd, Karar, Khukhuman, Barikarak, Sohdag, Athan, Lohra, Adhaura, Garale, Sarki, Urdag, Chapri, Kusumbha, Nagatoli, Rohtas, Rehal, Banda, Hasri, Budhua Kalan, Chakdih, Panda, Koras and Bholna.

5.7 ISOLATED RESIDUAL HILLS

There are many isolated residual hills (Plate : 1) completely detached from the main Plateau on the south-western, southern, eastern, and north-western margins of the study area. They are scattered and their altitude is low. They project themselves overlooking sky in the alluvial plain.

The southern isolated hills are small in height than those on the northern. Between the Son and the main plateau mass there are six isolated hills. They are Tiara, Parchha, Katudanr, Tenwa, Banua and Dumaria which rise upto 188, 198, 192, 165, 185 and 170 metres respectively. These **pahari** are made up of the Amarkha sandstone (Lahiri, 1964).

There are seven isolated residual hills in the northern part of the Plateau. They are **Khaimuri pahar** (287 m.), **Ghamha pahar** (298 to 301 m.), **Garda pahar** (325 m.), Bhagwanpur (276 m.), Ibrahimpur (233 m.) and Kalwari hill (336 m.).

All these hills appear originally to be part of the main Plateau mass which were subsequently detached from the main Plateau most probably due to collapse of caverns and further erosion. The northern, north-eastern and eastern isolated hills maintain more or less the same height of limestone and shale beds of fifth cycle of the Semri series (Table : 3 by S.M.

Mathur, GSI, Vol. 112 Pt. 8). These beds are overlain by the Kaimur sandstone. The **Murli pahar** is an interesting example showing two beds lying unconformably over each other in isolation (Plate : 1). As regards isolated hills in the south lying between 83^0 45' to 83^0 55' in the 63 P/14 maps just north of the Son river (Amrakha hill & others), they cannot be regarded as composed of the same Kaimur sandstone because of their low heights mostly below 200 metres. These hills appear to be of second or third cycle of the Semri series which has succession of sandstone over shale. The overlying beds of the Kaimur sandstones and Rohtas shale/limestone have been removed and denuded by large scale collapse of earlier caverns. A discussion with Shri T.N. Mishra, Asstt. Director, B.M.D.C. Ranchi, confirms this view.

TABLE : 5.2

DISTRIBUTION OF THE ISOLATED HILLS

Name of the mills	Location	Height in metres	Shape & Size
Tiara	24^0 32' - 83^0 41' 45"	188	Very small, conical
Parchha	24^0 32' - 83^0 43' 15"	198	Elongated east-west, like a small ridge
Katudanr	24^0 32' - 83^0 44' 30"	192	Conical with slight elongation in the east
Tenwa	24^0 33' - 83^0 46' 15"	165	Elongated east-west to north-east
Banua	24^0 33' - 83^0 47' 30"	185	Tapering towards north east
Dumaria	24^0 34' - 83^0 48' 15"	170	Roughly triangular
Murali	24^0 39' - 83^0 59' 00"	286	Conical, tapering towards south-west
Khaimuri	24^0 33' - 83^0 33' 45"	287	Glove shape with narrow portion facing east
Ghamha	24^0 56' - 24^0 58' 83^0 30' - 83^0 32'	301	Sword type running south-west to north- east
Garda	24^0 56' - 24^0 58' 83^0 31' - 83^0 34'	325	Makes animal like profile viewed above, with head towards the west & tail towards east
Bhagwanpur	24^0 56' - 24^0 58' 83^0 35' - 83^0 36'	276	Like english letter 'S' extended north tosouth
Ibrahimpur	24^0 52' - 24^0 58' 83^0 43' - 83^0 44'	233	Ridge like, extending east-west with wider area in the east
Kalwari	24^0 51' - 24^0 52' 83^0 43' - 83^0 44' 30"	336	North-south extending like ridge

5.8 CHUANS

Chuans are the small form of natural springs or waterholes working as a perennial source of drinking water over the Plateau which is dry and has acute crisis of water even for non-drinking purposes. Actually these **chuan** attracted the people in need to settle near them and consequently, there was a growth of small hamletted and scattered settlements in limited number with the least natural endowments for their growth in this dry, barren and inaccessible Plateau. In the old stratified sandstone, dug wells are few in number and even those dry up in the summer months due to a fall in water table. **Chuans** on the other hand supply water throughout the year because of the special geological stratification of sandstone and shale beds. Sandstone is pervious allowing the water to sink underground. Shale is impervious which help in maintaining a large reserves of underground water in overlying sandstone bed. Water seeps out throughout the year through these reserves and comes on the surface as spring where the shale beds reach close to the surface. The **chuans** do not dry up becsuse of the trapped large reserve of underground water. The flow of water in the summer decreases as the pressure of water becomes less. But they are fairly rapidly flowing streams in monsoon months and in the winter season . They supply clean naturally filtered water on which the people of the area depend. It is a pleasure to have a bath in the chuans. This is an important fascinating feature of the Rohtas Plateau.

The presence of **Chuans** over the Plateau has mainly two geomorphic deductions. Firstly, there is a bed of shale in between sandstone aboce and limestone below disallowing percolation of water into the limestone in areas close to the **chuans.** Secondly, there is a process of water percolation during the rainy season in large scale which goes underground into the limestone beds fascilitating the development of limestone caves and other associated features. The **chuans** are more or less perennial but there is no perennial. This shows that as soon as the spring water finds the bed of sandstone out of shale, the water starts percolating without any further continuance.

5.9 WATERFALL

Waterfall (Plates : 3&4) and rapids are found along the courses of almost all the rivers descending down from the Plateau, particularly towards the east and north directions. Some of them are very high and picturesque. The big waterfalls are mostly along rivers flowing towards the north. But even on the south where the streams are small the rainwater descends down in the form of cascades along the steep scarp-face. The author has personally watched the appearance of such cascades in monsoon months year after year from his house. The falls disappear in the hot summer season.

Like the waterfalls of Ranchi, Hazaribag and Palamu districts the waterfalls of this area also have originated due to faulting and associated block movements. The main difference, however, between the falls of the two regions is that while the waterfalls in other districts have been produced by faulting and uplift of the concerned areas, the falls in the Rohtas Plateau owe their origin to solution of limestone and collapse of limestone caves. The main Plateau is completely free from faulting but all the waterfalls are found at the top of the Plateau. 'The resistant sandstone of the Kaimur Plateau cause waterfalls in Shahabad' (Ahmad, 1965, p.36).

PLATE : 3 WATERFALL

A close view of Tutla Waterfall depicting the hard Vindhyan sandstone

PLATE : 4 PLUNGE POOLS & SUBSIDED VALLEY

Tutla Waterfallwith plunge-pool

The only explanation for the origin of waterfalls on this fault free surface is related to the geological structure. In most cases the overlying sandstone strata have collapsed and produced waterfalls and steep-sided valleys along the rivers.

The details of main waterfalls are given in table 5.3. The following are the characteristics of the above mentioned waterfall.

(i) There is complete structural control on these waterfalls. It is noted that the waterfalls of the adjoining highlands of Chotanagpur namely Hundru, Johna, Dasang etc. have been originated owing to the uplift of the earlier planated surfaces. The structure is granite and gneiss. Those waterfalls are subject to erosion and subsequently recession. They have been modified, sometimes vigorously. Therefore, the magnitude height and attractions of these waterfalls have been decreased. Some of those are not a single waterfall but by erosion and denudation there has been stepping along the river making a series of breaks along the river courses.

There is a different case with the waterfalls of the Rohtas Plateau. These are not due to uplift or differential erosion, but the caverns of the limestone beds below the sandstone beds have been mainly responsible for the developments of these waterfalls. The limestone is always soluble and the overlying bed of sandstone will always work as a strong capping. So there will never be modification in the magnitude, height and attraction of these waterfalls till the existence of the Plateau. The magnitude is such that the wall of waterfalls has a gradient of more than 90^0 or so because the sandstone capping is over hanging (Plates : 3 & 4).

(ii) In spite of the high magnitude and altitude of these waterfalls, they do not remain attractive throughout the year; because the streams flowing over them are not perennial. They do not survive even during winter months. They are absolutely rain-fed therefore, the attraction of these waterfalls lasts only for three or four months particularly during rainy season.

(iii) These waterfalls are caprock fall in nature.

(iv) There is every possibility of the formation of new waterfalls in the Plateau area if there is a further case of collapse of sandstone beds over the limestone caverns. The northern portion have much more possibilities.

(v) The steepness of the southern scarps is such that numerous waterfalls can be observed during the rain for only two or three hours. These are along the southern flowing small first order short-lived streams. After the rain is over there may not be any sign of stream course or waterfalls.

5.10 WATERKUNDS

Water **kunds** or pools are found at the foot of almost all the waterfalls of the Plateau. Some of them are of large areal extent while others are small depending upon the height of the waterfalls and volume and velocity of the falling water. 'Large falls regularly develop a considerably over-deepended enlarged during recession of the fall' (Fairbridge' 1968, p.1219).

TABLE : 5.3

WATERFALLS

Toposheet Nos.	Name of the Rivers	Long. & Lat.	Name of the Waterfalls	Height in m.
63 P/9	Sura I	83^0 32'-24^0 50'	-	80
	Sura II	83^0 36'-24^0 50'	-	130
	Karsotha	83^0 38'-24^0 46'	-	102
63 P/10	Phulwari	83^0 40'-24^0 35'	-	20
	Hasadah	83^0 59'-24^0 40'	-	30
	Bajra	83^0 58'-24^0 47'	-	50
63 P/14	Maspurwa nala	83^0 48'-24^0 44'	Sitakund fall	80
	Bahagawa nala	83^0 46'-24^0 43'	-	80
	Kalanwa	83^0 46'-24^0 40'	-	10
	Kalanwa next	83^0 46'-24^0 41'	-	30
	Gaighat nala	83^0 51'-24^0 39'	Rakimkund Fall	168
	Gurudan nala	83^0 51'-24^0 42'	-	40
	Gurudan nala	83^0 51'-24^0 43'	-	28
	Ausane Gurudannala	83^0 52'-24^0 42'	Kuaridah Fall	180
	" next	83^0 51'-$24^0$45'	Okheriankund Fall	90
72 D/1	Dhoba	84^0 02'-24^0 49'	Dhuadhar Fall	30

The origin is thus invariably related to the impact of the falling water which scours out the bed immediately below the falls since the debris falling with water also plays its part. The area of the pool gradually increases in size. The churning and drilling action of large boulders is responsible for the greater depths of some of these **kunds** (Plate : 4).

Although all the plunge pools are necessarily deep and cool, the plunge pools along the margins of the Rohtas Plateau have some peculiar characteristics of their own. This peculiarity is related to the geology of the area. In many cases, particularly the plunge pools of the Hundru, Johna, Dasang and other falls of the Chotanagpur highlands, the structure of the head of the waterfall is similar to that of the base. In case of the Rohtas Plateau there are varied horizontal strata of sandstone shale and limestone. The streams here flow over the old, red and resistant sandstone and descends down to the shale or limestone beds which are comparatively softer and soluble. The plunge pools in the Rohtas Plateau are, therefore, very deep. It is while these plunge pools have considerable depth of water. It is the case with almost all the plunge pools, e.g., Sitakund below the Sitalkund waterfall on the **Maspurwa nala** near Guptadham and **Dhuadhar kund** below the Dhuadhar waterfall on the Dhoba river near Tarachandi.

Another striking characteristics related to the magnitude of these plunge pool is their greater size. It has been possible not due to flowing and falling water of the waterfalls but owing to limestone. We have noted earlier that the waterfalls have been originated because of the collapse of material over the caverns. So is the case with the formation of these large plunge pools. These are not confined only to the waterfall zone; the pools extend to their greater width (Plate : 4).

Most of the plunge pools of the Rohtas Plateau are distributed on the north face bounded by high vertical rocky walls on the east and west. Therefore, these plunge pools get sun rays neither in the morning nor in the afternoon. These are extremely cool so much so that few can dare to take bathe in these pools even in summer. It is the case with all the seventeen **kunds** observed by the author.

5.11 APPLE SHAPED VALLEYS

The Rohtas Plateau has peculiarity of having the apple shaped valleys which are the product of heterogeneous vertical structure and Karstification. This feature looks like an apple with small circumference on the top and larger circumference in the middle and at the bottom. This is virtually and aggregate expression of waterfall, plunge pool and subsided valleys. But the shape is quite different from normal feature of waterfall. In other waterfalls too plunge-pools develop at the bottom but these are not encircled by high-walls around except an out-let which also forms a steep-sided U shaped valley. Such features are very much pronounced at Sitalkund waterfall (83° 48'-24° 44'), Sura I (83° 32'-24° 50'), Sura II (33° 36'-24° 50'), Karsotha (83° 38'-24° 46') etc.

As we have noted earlier, waterfalls of the Rohtas Plateau owe their origin neither to uplift of planation surfaces at different height as we find in most cases of the Chotanagpur Plateau, nor the differential erosion caused by horizontal inequivalence of structure. These are essentially due to variation in vertical structure which consists of hard sandstone, shale and limestone. It is also noted that the valleys are deep not due to gradual process of fluvial erosion but the above noted characteristics of the valleys reveal that they are undoubtetly subsided. This also reveals that there was a Karstification in the geological past below the thick capping of the Vindhyan red sandstone. The hollowness and the subsequent earth-movement have been responsible for the development of the apple-shaped valleys.

5.12 SUBSIDED VALLEYS

The lower reaches of the rivers which flow below the waterfalls but considered to be a part of the Rohtas Plateau are characterised by vertical valley-sides, sometimes more than 90°, and undulating but wide valley beds consisting of sandstone boulders, calcite and shale materials. Such valleys are but the developed form of notches and **khohs,** discussed earlier. The occurrence of sandstone boulders in large scale, steepness of the valley-sides and the presence of limestone strata below the sandstone capping all reveal that these valleys are not due to normal erosion processes but they owe their origin firstly to karstification in the forms of caverns and later to the subsidence caused by earth-movements particularly during the Tertiary-era.

Such valleys are pronounced along the river Sura I, Sura II, Karsotha, Ausane, Goptha etc.

5.13 CAVES & THE ASSOCIATED FEATURES

Among the features which are fascinating and significant in the study of the landforms of the Rohtas Plateau and which are not easily accessible and observable to casual passers by are the caves (Plates : 5 & 6) and the features developed therein like stalactite, stalagmite, dripstone, vadose and phreatic streams and other depositional features both allochthonous and autochthonous. The features which have been discussed previously are open to casual observers but these features are hidden in caves which requires enthusiasm and academic persuites for their study. Till the date such features were not studied, rather not came into light except the Gupteshwar cave. It may be because of the fact that features of early stages in a karst region like sink-holes, dolines, uvalas etc. are not developed on the surface owing to the 300 feet thick capping of the old hard Kaimur sandstone but the percolation of water has facilitated the development of caves and the associated features. These features are among the best examples of karst landforms in India. The author has tried, in spite of many odds, to explore the features. He is, however, humble to accept that many areas are left still unexplored. Here are the descriptions and the possible analysis of the origin of some of the features.

CAVES

The oxford dictionary defines a cave as 'an underground hollows usually with a horizontal opening'. Caves, however, are not found necessarily with a horizontal openings. 'In terms of human experience we generally think of a cave as being a natural roofed cavity in rock which may be penetrated for an appreciable distance' (Sweeting, 1972, p.130). Caves have different connotations in different country. In Britain the terms is used to include both horizontal cavities and vertical passages. In other countries there is a different word for vertical passages. We shall however, refer to all kinds of underground hollows under this general terms.

Caves are generally found in area composed of limestone the world over. Since Rohtas Plateau is also such an area, caves have formed in many parts which have not been fully investigated till now. Past researchers have not been able to identify the major caves except a few like Gupteshwar cave on the northern scarp face. The Gupteshwar cave came to lime light on account of its status as a centre for pilgrimage where annual fair is organised on the occasion of Shiva-ratri generally in the month of February.

The description of the Gupteshwar cave is available, therefore, in the District Gazetteer in accepted details; and it is sorry to state that no further addition was made to the knowledge in subsequent researches and writings; it has just been a reproduction in the form of quotation or changed version. It was not even bothered to see whether caves other than the Gupteshwar were existing in the Plateau area in spite of the acceptance that the whole plateau has underlying limestone beds and the area is falling under sub-tropical humid region where precipitation and percolation is natural. The fact is that there are numerous caves in the study area peeping to the Son river on the south and east and the Gangetic plain on the north either through the narrow hole or through the wider mouth. The caves on the northern and northeastern margin are, however, more prominent and apparent than those on the southern margin. It is but the natural because the area is sloping towards north and east facilitating the flow of the surface an

PLATE : 5 LIMESTONE CAVES

Outer view of the Chunhatta Caves

Plate - 5

PLATE : 6 LIMESTONE CAVES

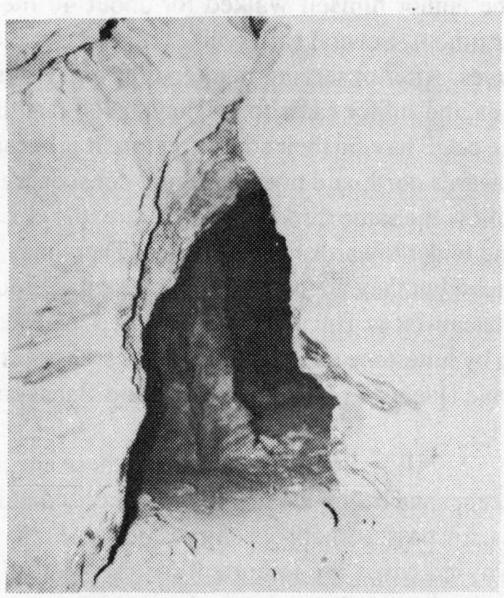

Inner view of the Chunhatta Caves

Plate - 6

TABLE : 5.4

DISTRIBUTION OF CAVES

Name	Location	Dry/Wet	Features
Bansi	83^0 29' 30" 24^0 34' 15"	wet	Small caves
Chunhatta	24^0 36' 83^0 52'	wet	Only dripstone came into existence in limestone quarries
Banjari	24^0 41' 83^0 59'	dry	Limestone features are not developed in this cave
Karkatgarh	24^0 53' 83^0 22'	wet	Small cave
Gupteshwar	24^0 45' 40"	wet	Stalactites, Stalagmites, Dripstone

underground drainage towards these directions, and consequently the development of the mouth of the caves on the north and east. The caves on the southern bold unindented scarps are, however, not absent, may there be their source area or the less developed. The author has traversed the area during his field work and has been able to locate two fully develeped caves in the southern part. They are Bansi cave on the south-western part of the scarp and the Banjari cave on the eastern side (Table :5.4). The caves were penetrated and they seemed to extend far under the Plateau. The author himself walked for about 40 metres inside the caves but was unable to trace its terminous. Several gallery like branches seemed penetrating both on the left and right side of the caves. After observing many points and making enquiry from the persons acquainted with the area, the author came to the conclusion that the southern and eastern Plateau faces have mostly the caves having their source region. It is because of the fact that the Plateau in general is sloping towards north and north-east and consequently the surface streams and underground drainage follow the same direction. Therefore, the caves are narrow on the south and they become wider and wider towards north and east. There may be many caves on the south and south-east scarps faces, but they are not open; because their source may be a little further from the escarpment or Plateau-faces. But wherever the scarps have receded northward or eastward either by denudation or by limestone quarries; the source area becomes open to the observer, as is the case with Banjari cave (Plate : 6) near Kalyanpur and Bansi cave near Amhuan (Table : 5.4).

It is, however, noted that these caves are extending towards north downward, they go deeper and become wider and channelised. It is difficult to enter into these caves because of their narrowness; sometimes entry is possible only by creeping which is risky and dangerous. This is why no caves are important for the visitors on the southern and eastern faces of the Plateau excepting the Banjari cave into which persons enters with risk but religious sentiment on the occasion of Makar Sankranti (14th Januaray) to offer puja to Lord Shiva and Goddess Parvati by travelling and creeping for about 30-40 metres. The caves have developed stalactite and stalagmite which are believed to be God and Goddess.

The author also feels that on the southern part of the Plateau the caves are there extending south-north and having wider parts underground towards north. These caves are, however, not open. They will be open only after the large recession of Plateau in future. As regards the eastern and northeastern portion, the caves are limited in number, because the limestone beds have no considerable height, sometimes they project into the interior towards north below the son river and alluvial deposits. So, there is less scope for the developement of underground channel-facilitating the development of caves. These are the areas of water seepage. In Banjari areas, even limestone quarries are full of water and limestone is extracted after water is pumped out. All these are to say that the existence of caves in the eastern and north-eastern portions of the Plateau is limited.

The northern scarp faces, however, possess caves of different magnitude. In this part the caves have their mouth, so one can enter into the caves and can travel as long as they become narrow and intricate or dark with gaseous suffocation. Almost all the notches which are supposed to be the former caverns possess some points of entrance into the caves. Of the caves located on the northern margin, the Gupteshwar cave is the most important. There are many other caves along the valleys of Sura I, Sura II, Karsotha etc. rivers. But they have not been fully explored. The main reasons for their remaining incognito are two fold. Firstly the caves occur in dense forest areas where communication is not easy. Secondly the environment around these caves is solitary and manage to reach the caves without adequate security aids. There are no villages in the vicinity of a caves and there is no human activity whatsoever. Definite information is available about a cave along the right bank of the Karamnasha river about 1.6 kilometres of Karkatgarh. The cave has been found to contain some ancient paintings on the ceiling and walls. The matter has been reported in the news bulletin of G.S.I. (Vol. 7. No.1, 1976, p.13).

Caves in limestone originate due to horizontal bedding and vertical joints which provide passage for rainwater changed with CO_2. As long as water keeps flowing through joints the solution of limestone continues and the enlargement of the passage keeps pace with this process. After underground hollows of appreciable size develop the caves are enlarged by the additional processes of gravity sliding and slumping. Since the capping of the Plateau consists of hard sandstone the process of cave formation has become easier because the hard sandstones do not subside very easily and continue to remain as cave roofs for a very long time. In the meantime the underlying limestone keeps dissolving and flowing with the underground passage of water. Therefore, the caves of the Rohtas Plateau are very long and there is mystery about their total length. No body has been able to find out the total length of the underground tunnels either in the Gupteshwar cave or in other caves. Their huge dimension exists due to the protection provided by hard sandstones at the top.

Owing to the general tilt of the Plateau from south to north, surface water flows from south to north. Similarly underground water also flows from south to north. Therefore, the solution processes work faster in the northern part of the Plateau and consequently the caves of the north are longer and larger. Solution is not so rapid in the southern part because of low pressure of water and there has been development of natural caves only at a few places. Moreover, during dry months the water table sinks down and southern high scarps became almost waterless whereas in the northern half water is available almost throughout the year. The process of solution and cave widening is a continuous process in the north and intermittent process in the south.

The author has done detailed study of Gupteshwar cave which is easily accessible during festive occasions. The dimensions and the plan of the cave are described below :

GUPTESHWAR CAVE

The cave of Gupteshwar are situated in a narrow precipitous glen in the Rohtas Plateau, about 11 kilometres from Shergarh. The entrance lies a little way up the hill, and is about eighteen feet wide and twelve feet high. The bottom, roof and the sides are rough and rocky. At some places it has steep ascends and descends and at the other the floor slopes down uniclinally making the path narrower and rocky. So, walking inside the cave is difficult almost everywhere.

The cave extends first towards east from the mouth for about 380 feet. It terminates in a low part called **'Patalganga'**. At the end is a smaller hole into which no body can enter, but it seems that it continuous further to great distance (Fig. 8).

The cave has an off-shoot known as **'Koonjgali'** (Narrow lane) which runs from its mid point to the south-east up to about 87 feet and rejoins the first gallery (Fig.8). Further there are two narrow ways to enter into the next gallery. The one being lower and the narrowest passage forces visitors to travel by creeping on their hands and knees up to about 11 feet. The other one is not so lower and narrow. The former is known as **'Brahamyoni'** and the latter **'Sumbha'.** Both lead to the other gallery extending eastward (Fig.8). This is 370 feet long and is comparatively wider than the **'Kunjgali'.** Near its mouth beautiful stalagmites have developed on the floor. The stalagmites are regarded as replicas of **'Panch Pandavas".** Hence, this is also called as the **'Panch Pandava cave'**. A little away from this feature is the accumulation stalactites hanging from the ceiling.

At a distance of about 140 feet from **'Panch Pandavas'** eastward extension of this cave is locally called **'Ghordaura'.** Here it is lofty and wide. It has one sided slope. The surface is marked by erosional pits resembling the hoof marks of horse. The surface is silty and very slippery. Further east it shrinks in size. Erratic speleothems have developed here. Helectites are most common which extend from vertical cave wall and mini stalactites. After the **'Patalganga'** point the cave becomes constricted and so remains unexplored.

The northward passage of 92 feet long from the point of cave crossing leads to an apartment called **'Tulsi chaura'** or **'Nachghar'.** The central part is wide and lofty and resembles with a hall. The ceiling is quite plain. The term **'Tulsi chaura'** denotes a platform on which the sacred **'Basil'** is grown but this place is quite devoid of **Tulsi** plants.

The main **Gupta** cave extends towards south, 240 feet in length. It gradually becomes wider. It is widest where the fanciful stalagmite resembling **'Sivalinga'** has developed on the floor. This is the main object of worship. From the ceiling are suspended stalactites with dripping water. These are regarded as the matted lock of the lord. Pilgrims try to have a drop of water in their open mouths falling from the ceiling. The floor of the cave is wet all through. At places water accumulations are found. A bit ahead, the cave has two arms extending in the opposite directions. Being narrow and dark they remain unexplored.

DEPOSITIONAL FEATURES IN CAVES

Depositional features found in caves have always fascinated geomorphologists. The features found in limestone caves do not occur in areas composed of other types of rocks. Therefore, particular attention has been paid on a study of depositional features in limestone caves throughout the world. The feautres found in Rohtas caves have been mentioned by other researchers but most of them are associated with the features found in Gupteshwar cave.

The depositional features found in the caves are not common in all the caves of the area. Normally the depositional features have been divided into two parts : (1) allochthonous features which develop on account of material brought from the caves and (2) autochthonous features which develop within the cave due to various processes. In the first category deposits of clay terra rossa soil etc. are treated which are generally found on the floor of the caves under flowing water. These allochthonous features of clay deposits are found in almost all the caves because all of them contain underground stream flow on the cave floor. In some cases sand deposits have also been observed which were deposited after the disintegration of upper sandstone beds. The upper sandstones are of a red tint consequently the deposits are also reddish in colour resembling with terra rossa soil. The clay deposits have been caused by decomposition of shale strata which sometimes form the floor of the caves and sometimes a portion of the cave walls. The joints in the limestone allowed the passage of clay solution which are released on the floor to be deposited as thin layers. Since the flow of water is very slow the clay gets enough time to settle down on the floor. In a fast flowing water they are carried away in suspension.

The autochthonous features include stalactite, stalagmite, dripstones, curtain, helictites etc. for which a general and widely accepted term in geomorphology is speleothems.

The most interesting and outstanding geomorphic features inside the cave are depositional features of cave travertine mainly associated with limetone topography. In the Gupteshwar cave **panch pandavas** are beautiful stalagmite of different sizes. Above them there is a large number of stalactites hanging from the ceiling with constantly dripping water. Another fanciful stalagmite which has in the form of **Shivalinga** is about 4 1/2 feet high. It is regarded as Lord Shiva. The head of this is really a typical depositional feature developed in the limestone cavern.

Stalactites in every cave have been caused by the dripping of water from cave roofs. 'Percolating water enters caves through cracks and joints in the roof and loses its excess CO_2 on mixing with the cave atmosphers. As the water drops enter the cave the CO_2 is lost, calcite is deposited as a tiny ring where the surface of the drop is in contact with the cave roof. As more drops emerge, ring upon ring of calcite is added to form a hollw cylinder, the diameter of which is the same as the diameter of the drops of water; the average size of a water drop is 5 mm. As the cylinder grows the water moves down through the central hollw and adds a small increment of length to the stalactite' (Sweeting, 1972, p.178). In this way stalactites have been formed in the caves of the Rohtas Plateau. The stalactites vary in size from about 5-10 mm. in diameter to

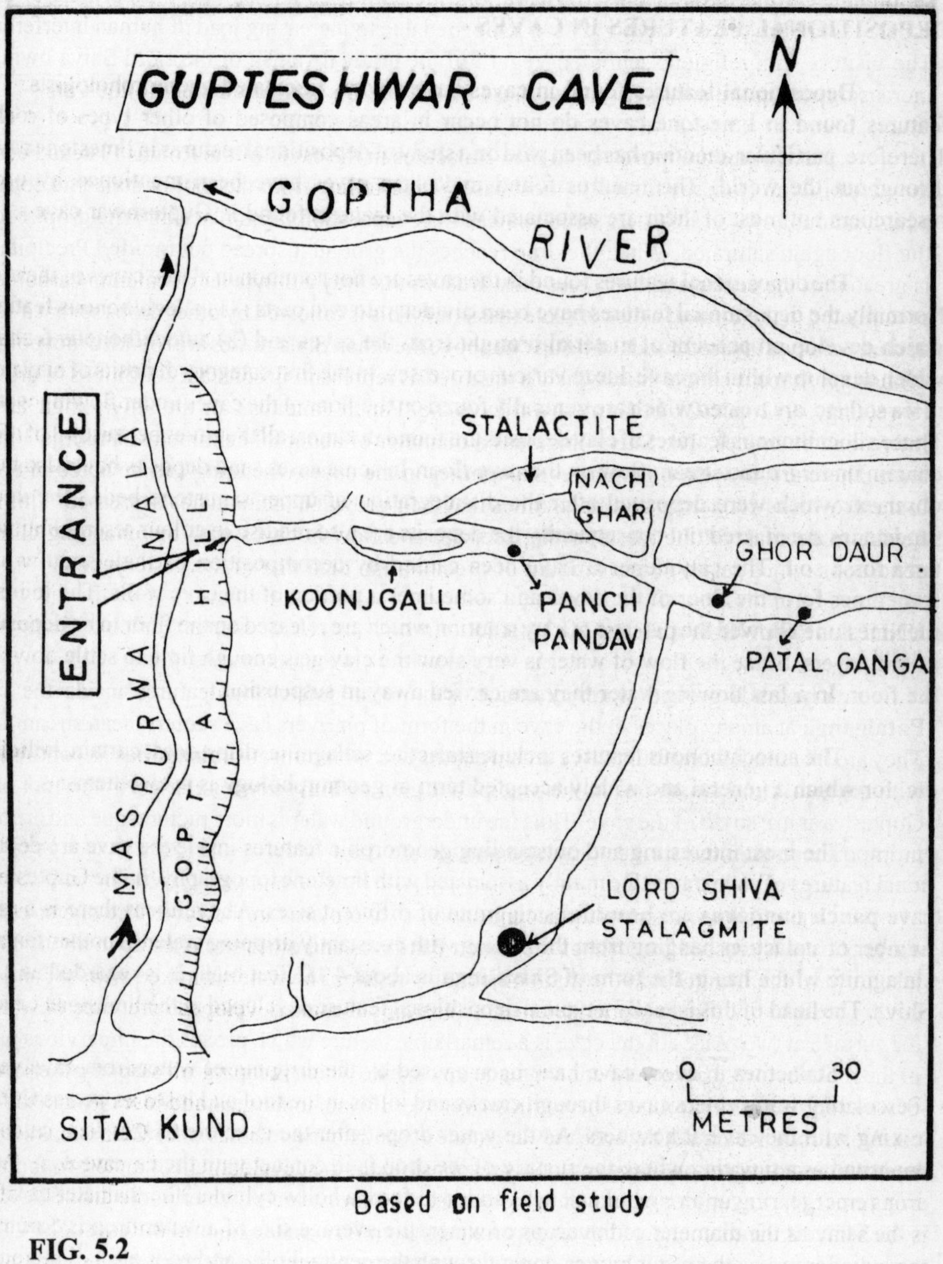

Based on field study

FIG. 5.2

several metres long. In the Gupteshwar cave they hang up to 3 metres in accumulation which look like bearded roof. These stalactites are regarded as beard of Lord Shiva by visitors with religious sentiment. On the basis of close observation, the author could see that the portions of these stalactites are broken which might have happened due to increasing load of human interference. The visitors with religious sentiment regard of the gradual vanish of the Lord Shiva owing to increasing sin on the earth. Any way, these stalactites are very fascinating to look at.

Stalagmites are more varied than stalactites in the caves of the Rohtas Plateau because they are built up from the floor. 'In most cases water falling from the roof or stalactite contains a high amount of Ca $(HCO_3)_2$; as the water falls evaporation takes place and the drop reaches the floor again saturated. When the drop reaches the ground it spread out rapidly. Precipitation is greatest at the centre and decreases outwards. In this way a thin layer of calcium carbonate is precipitated in the form of a disc or plate (assiette). This is followed by other drops so that a series of dome-shaped layered hoods build up one above the other. Stalagmites are rounded, not pointed; the normal shape is candle-like, which is regarded as representing constant conditions of growth' (Sweeting, 1972, p.182). In Gupteshwar cave the 'Panch Pandavas' or the Shiva-linga, are the interesting examples. The Shiva-linga has a diameter of about 80 centimetres and it is 3 meter high. There are, however, numerous stalagmites in the caves.One thing can be noted in regard to the growth of stalactites and stalagmites that their growth in general has decreased. This may be due to decrease in the rate of water seepage and flow in caves with the changing climatic conditions which result in less amount of precipitation than that of the yesteryerars.

UNDERGROUND WATER FEATURES

The underground water drainage is quite an interesting features inside the cave. **Patalganga** at amany places in the cave in the form of pit rivers have subterranean stream flow. They are like small wells four to five feet in depth. The subterranean drainage has its continuation with the **Sitalkund** which lies in the precipitous glen of **Maspurwa nala** just at the foot of the Gupteshwar hill south of the cave. Thus the underground water is more picturesque and is no less in importance than the surface drainage outside the cave.

There is knee deep water accumulation at two places in the cave. The one stretching about 50 feet is crossed just east of the Panch Pandavas and other stretches up to about 80 feet just north of the Lord Shiva, the main object of worship. The water is mixed with silt and clay having reddish tint due to preponderance of ferruginous matter. Thus, the presence of water of the surface at two places in the cave is a remarkable feature which proves the impervious nature of the rocks below. The lower beds are made up of shale which is capable of stopping the seepage of water further underground. There is however, free flow of underground water through the cave roofs or through the walls because they are made of sandstone and limestone. Continuous seepage and dripping of water on the cave floor is responsible for the development of underground water feautures. The cave floor is uneven which causes the formation of water pools inside the caves which have are interconnected by underground stream flow on the floor of the cave. The visitors are required to wade through knee deep water at many places on a sleepery surface passage in the caves in troublesome and dangerous due to these underground water features.

6

RELIEF MEASURES

Relief measures of a region are interrelated with slope. Both slope angle and sediment loss (hence rate of change of slope) are the greatest where relief is at a maximum. The maximum Plateau Relief is 615 metres but the extensive areas lie between the 300 m. and 500 m. contours, giving a general relief of 300-500 metres for the most of the area (Fig. 5.1). The high relief in the Plateau may be associated with areas of tectonic activity.

Since relief measures are indicative of the potential energy of a drainage system, by virtue of elevation above a given datum a detailed discussion has been made by using several methods of relief representation, such as relative relief, absolute relief, average relief or intensity of relief, dissection index, ruggedness index, and their cross-correlations. For the determination of these relief aspects, 1:50000 Topographical sheets (63 P/5, 6, 9, 10, 13, 14; 72 D/1, 2) have been divided into two centimetres grids each representating one squqre kilometre, and then maximum and minimum heights in each grid have been noted to obtain values for 2340 grids. With the help of these values, the relief aspects* have been discussed which undoubtedly correlate with slopes of the Rohtas Plateau.

The objective of this study is also to determine and critically examine if the Davisian Scheme (1909, pp. 249-278) of classification of landform denudation by stage has any quantitative implications, or if the scheme can be quantitatively described.

RELATIVE RELIEF

'Relative Relief' is one of the various methods evolved to measure the average slope. The term was invented and used by G.M. Smith (1935, pp. 275-284) to ascertain the amplitude of available relief to relate the altitude of the highest and the lowest points of any particular area. 'Lines of equal relief, isopleths, provide a somewhat better idea of relief than an ordinary contour map, since they are a measure of general steepness, though not of local steepness' (Fairbridge, 1968, p. 545). Thus, the Relative Relief map of the Rohtas Plateau (Fig. 6.1) shows somewhat more strikingly the areas of greatest local relief than does not any other ordinary topographic map. 'It also depicts the relief of an area in relation to the surrounding areas; probably so, it is called relative' (Prasad, 1985, p. 201).

A comparison of relative relief map (Fig. 6.1) with the surface elevation map (Fig. 5.1) of the Plateau clarifies this statement.

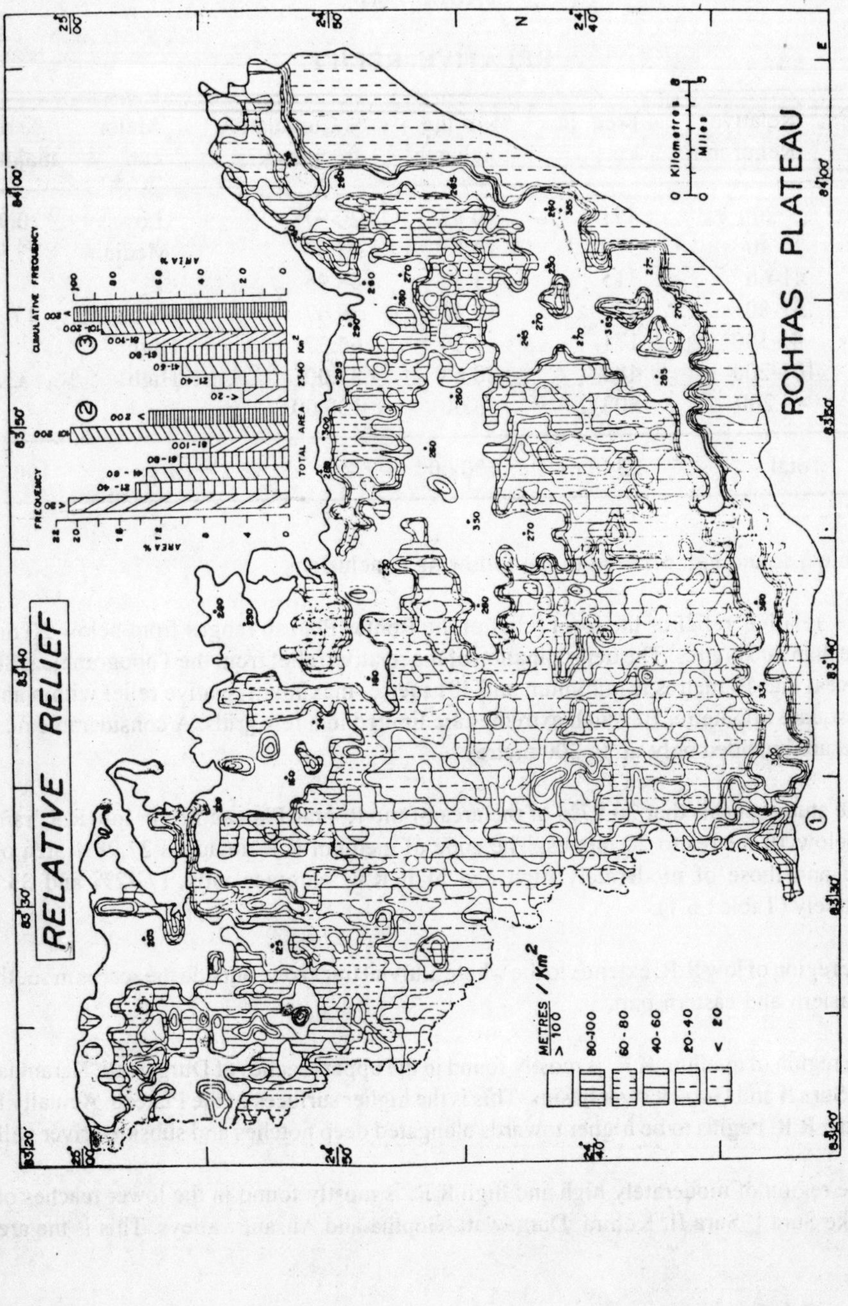
RELATIVE RELIEF
ROHTAS PLATEAU
FREQUENCY
CUMULATIVE FREQUENCY
AREA %
TOTAL AREA 2340 Km²
METRES / Km²
> 100
80-100
60-80
40-60
20-40
< 20
Kilometres
Miles

FIG. 6.1

On the basis of the data, extracted from the Topographical Sheets (1:50000), Relative Relief map (Fig. 6.1) has been prepared, and histograms showing percentage frequency and cumulative percentage frequency have been drawn (Figs. 6.1.2 & 6.1.3). These have been helpful in categorising the area, as noted in the Table 6.1.

TABLE : 6.1

RELATIVE RELIEF

No. of Cat.	Relative Relief in m.	Freq. in kms^2	% of the total area	% Cumulative frequency	Major cat.	% of the major cat.
I	< 20	491	20.97	20.97	Low	20.97
II	21-40	340	14.53	35.50	Medium	27.98
III	41-60	315	13.45	48.95		
IV	61-80	237	10.12	59.07	Moderately	17.42
V	81-100	171	7.30	66.37	High	
VI	101-200	483	20.63	87.00	High	33.63
VII	> - 200	303	13.00	100.00		
	Total	2340	100.00			100.00

The table 6.1, and Fig. 6.1 lead to the following conclusions :

1) The relative relief or the local relief of the Rohtas Plateau ranges from below 20 metres to more than 200 metres. The determination of the relative relief from the Topographical Sheets also reveal that 10 metres is minimum and 401 metres maximum relative relief within an area of one square kilometre. But such extremes are limited in a few grids. A considerable number of grids are included only up to 200 metres.

2) The study reveals that 20.97% of the area of the Rohtas Plateau lies in low R.R. ranging from below 10 metres to 20 metres. The zone of medium R.R. includes 27.98% area of the Plateau and those of moderately high and high R.R. embrace only 17.42% and 33.63% respectively (Table : 6.1).

3) The region of low R.R. extends to the whole alluvial tract zone outside the scarps in southern, southeastern and eastern part.

4) The region of medium R.R. is mostly found in the upper reaches of Durgawati, Karamnasha, Sura I, Sura II and Gurwat river basins. This is the higher surfaces of the Plateau. Actually from this zone, R.R. begins to be higher towards elongated deep notches and subsided river valleys.

5) The region of moderately high and high R.R. is mostly found in the lower reaches of the river like Sura I, Sura II, Kohira, Durgawati. Goptha and Ausane valleys. This is the area of

southern, southeastern and eastern scarps zone. In the northern scarps and mostly in all river valleys which is narrow, elongated, and gorge like, the R.R. is moderately high and high.

6.2 ABSOLUTE RELIEF

We have noted earlier that absolute relief is the maximum elevation of an area. The maximum elevation for each grid (one km2) has been extracted from the 1 : 50000 Topographical Sheets, and the Absolute Relief map (Fig. 6.2) has been prepared.

'The objective of this study is to determine how much erosion has taken place in relation to the present summits or hill-tops of the Plateau, because hill-tops or summits are generally the last vestige of vanishing relief' (Prasad, 1985, p.202).

TABLE : 6.2

ABSOLUTE RELIEF

No. of Cat.	Absolute Relief in m.	Freq.in sq. kms.	% of the total area	% Cumulative freq.	Major cat.	% of the major cat.
I	100-200	301	12.86	12.86	Low	21.70
II	201-300	207	8.84	21.70		
III	301-400	611	26.11	47.81	Medium	26.11
IV	401-500	757	32.36	80.17	Moderately	32.36
V	501-600	432	18.46	93.63	High	
VI	> 600	32	1.37	100.00		
	TOTAL	2340	100.00			100.00

The Table 6.2 and Fig 6.2 reveal the following points with noting.

The absolute relief of the Rohtas Plateau ranges from 200 metres to more than 600 metres in general. But the determination of the absolute relief from the 1 : 50000 Topo-Sheets also reveal that 110 metres is minimum and 657 metres is maximum within an area of one square kilometre. But such extremes are limited in a few grids. Only 210 grids fall in A.R. of 100 metres to 150 metres, 91 in 151 metres to 200 metres, 93 in 201 to 250 metres, 114 in 251 to 300 metres, 321 in 301 to 350 metres, 290 in 351 to 400 metres, 457 in 401 to 500 metres and 432 in 501 to 600 metres. Similarly only 32 grids have A.R. of more than 600 metres. A considerable number of grids are included between 300 metres and 500 metres. It covers 58.47 % of the total area of the Plateau approximately.

Thus the 21.70 % of the area lies in low A.R. ranging from 100 metres to 300 metres. The zone of medium A.R. includes 26.11 % area of the Plateau and those of moderately high, and high A.R. embrace 32.36 % and 19.83 % respectively (Table : 6.2).

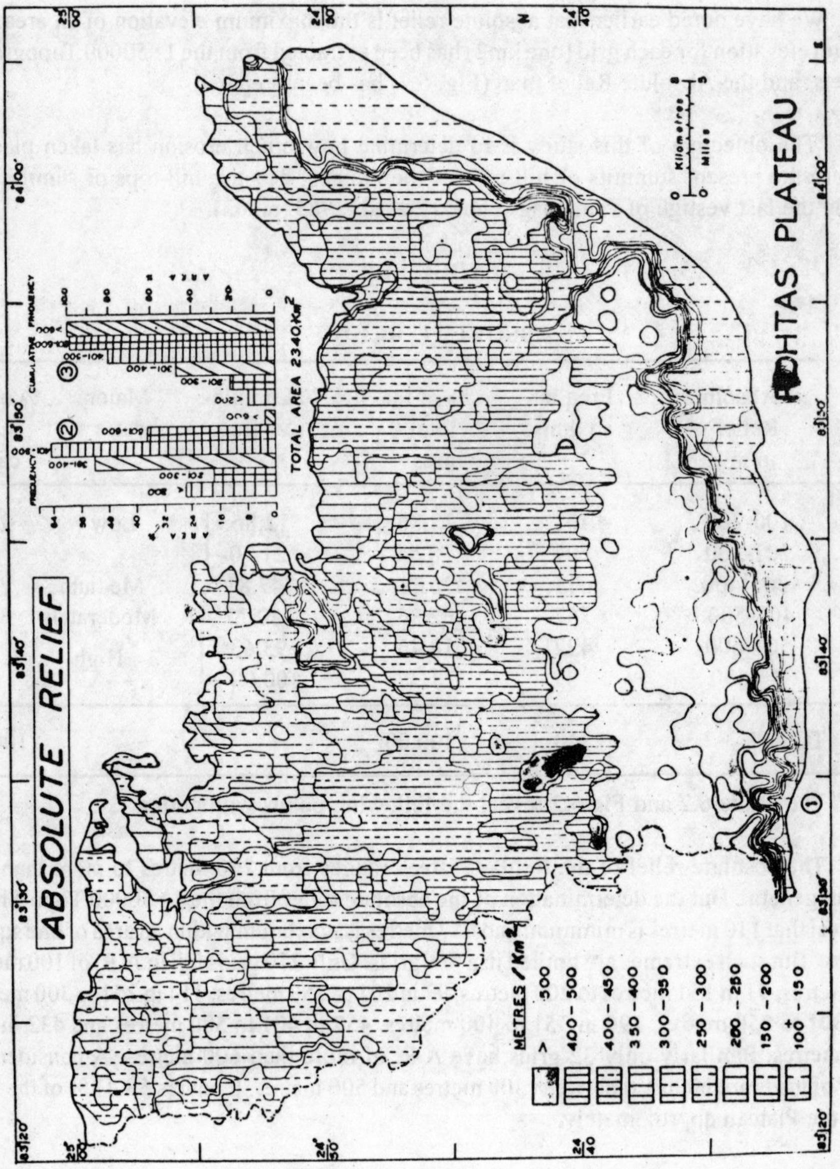
ABSOLUTE RELIEF
ROHTAS PLATEAU
TOTAL AREA 2340 Km²
FREQUENCY
CUMULATIVE FREQUENCY
AREA
METRES / Km²
> 500
400 – 500
400 – 450
350 – 400
300 – 350
250 – 300
200 – 250
150 – 200
100 – 150
Kilometres
Miles
83°20'
83°30'
83°40'
83°50'
84°00'
25°00'
24°50'
24°40'

FIG. 6.2

6.3 AVERAGE RELIEF OR INTENSITY OF RELIEF

Average relief is the expression of intensity of a particular area. It is called, therefore, intensity of relief also. It is an average or the mean value of the sum of maximum and minimum elevations. Therefore, it is somewhat different from the general contour map, **i.e.**, map showing elevations with the help of surveyed contours. The latter does not show relief, rather it projects the surfaces of varying elevations. On the other hand, the former, **i.e.**, the average relief depicts and categorizes the relief zones. Thus it is nearer to the relative relief of an area.

'The objective of this study is to determine how much is dissected the Plateau' (Prasad, 1985, p.204). For the purpose, data have been extracted from the 1 : 50000 Toposheets for 2340 grids separately (Table : 6.3) to prepare the Average Relief map (Fig. 6.3) and histograms (Fig. 6.3.2 & 6.3.3).

The Table : 6.3, and Fig. 6.3 give the following expressions:

1) The average relief of the Rohtas Plateau ranges from 100 metres to more than 600 metres. But most of the areas (42.34%) lies in low average relief or intensity of relief (I.R.) ranging below 200 metres to 300 metres. The zone of medium I.R. embraces 29.75 % area of the Plateau whereas those of moderately high and high I.R. include only 19.81% and 8.10 % respectively (Table : 6.3).

The zone of high I.R. is limited only to 190 grids (Each grid of 1 km2) comprising 5 grids of above 606 m. high, 90 between 551-600 m. and 95 grids in 501-550 m. I.R. zones.

TABLE : 6.3

AVERAGE RELIEF

No. of Cat.	Average Relief in m.	Freq. in sq. kms.	%of the total area	%Cummulative frq.	Major Cat.	%of major Cat.
I	100-200	457	19.52	19.52	Low	42.34
II	201-300	534	22.82	42.34		
III	301-400	696	29.75	72.09	Medium	29.75
IV	401-500	463	19.81	91.90	Moderately	19.81
V	501-600	185	7.89	99.79		
VI	> 600	5	0.21	100.00	High	8.10
TOTAL		2340	100.00			100.00

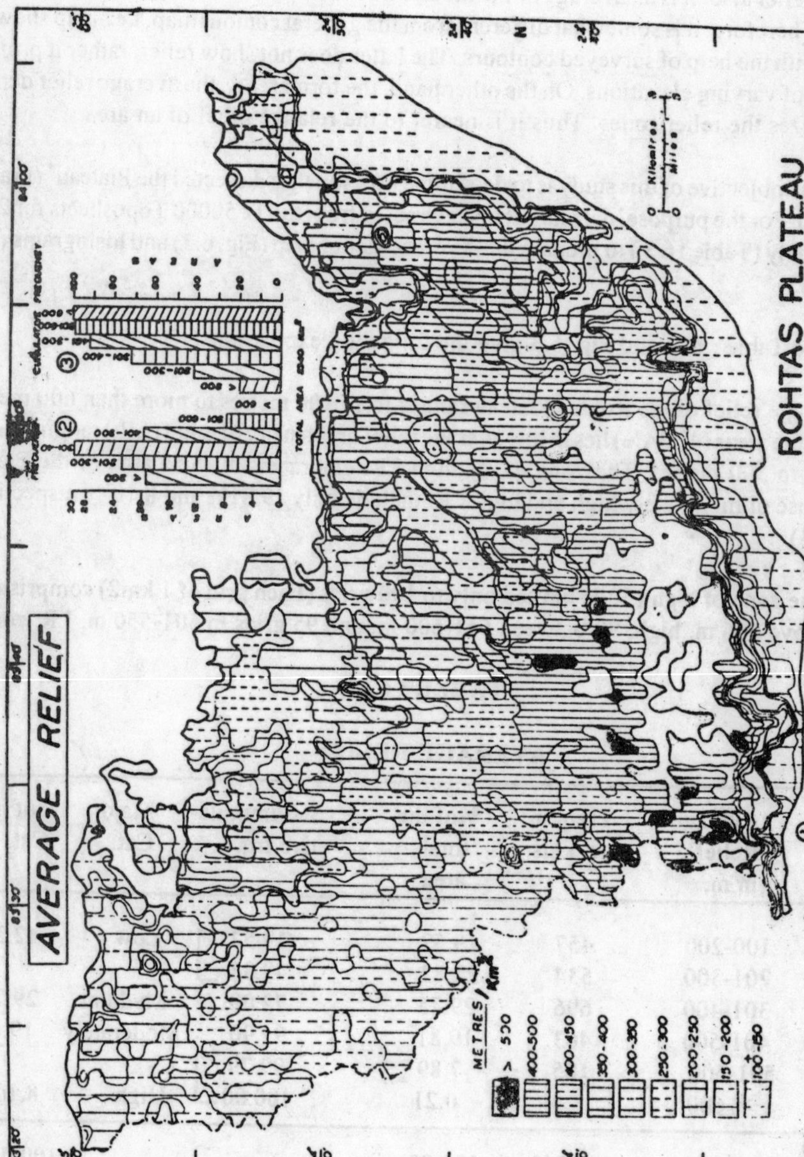
AVERAGE RELIEF
ROHTAS PLATEAU
METRES / Km²
TOTAL AREA
CUMULATIVE FREQUENCY
FREQUENCY

FIG. 6.3

6.4 DISSECTION INDEX

'An analysis of the dissection index of a region gives clue to the development of landforms under the perview of fluvial geomorphic cycle of erosion' (Prasad, 1985, p.206). It is the true index of the degree of dissection, and therby it clearly distinguishes the areas of high degree of dissection from low or medium or moderate or the like. And then it makes easy to infer whether a particular area is in youth or maturity or in old. Further it is also an indicative of slope regions of steep, moderate, or gentle, **etc.**

The study of dissection index (D.I.) has an extra merit besides that of physiography in the sense that D.I. is related not only to the elevations, but also to relief, dissection and slopes. 'It means that the area of lower elevation may be characterized by high degree of dissection index, if the area is dissected by deep river-valleys or is characterized by frequent and isolated hill-tops' (Prasad, 1985, p.207). Thus the study is near to that of relative relief. But it is not the same . The relative relief may indicate equal value for the higher surface (Ranging from 500 m. to 600m., R.R. being above 100 m.) and the lower surface (ranging from 200 m. to 300 m., R.R. being 100 m.). But the dissection index varies very much in such circumstances the higher surface amounting to about 0.17 - D.I. whereas the latter amount to about 0.33. It means that the bottom of the higher surface is far from the base-level whereas in the latter case, it is very near to the base-level. Where D.I. is 1, it is on base-level , and thus it is maximum. Therefore, the study of D.I. is essential to elucidate how much dissection has occurred on the landscape, and whether the denudational agents are nearer to goal or are far therefrom.

For the purpose, data have been obtained after dividing relative relief by absolute relief for 2340 grids and then maps (Fig. 6.4) and diagrams (Fig. 6.4.1. & 6.4.2) have been prepared. The Table : 6.4 has been helpful in categorizing the Plateau into different zones.

The Table : 6.4, and Fig.6.4, 6.4.1 and 6.4.2 lead to the following expressions.

TABLE : 6.4

DISSECTION INDEX

No. of Cat.	Dissection index	Freq.in metres	%of the total area	%of Cumu-lative freq.	Major Cat.	%of major Cat.
I	< 0.10	635	27.14	27.14	Low	27.14
II	0.11-0.20	613	26.20	53.34	Medium	38.20
III	0.21-0.30	281	12.00	65.34		
IV	0.31-0.40	183	8.25	73.59	Moderately	14.92
V	0.41-0.50	156	6.67	80.26	High	
VI	0.51-0.60	184	7.87	88.30	High	19.74
VII	> 0.60	278	11.87	100.00		
TOTAL		2340	100.00			100.00

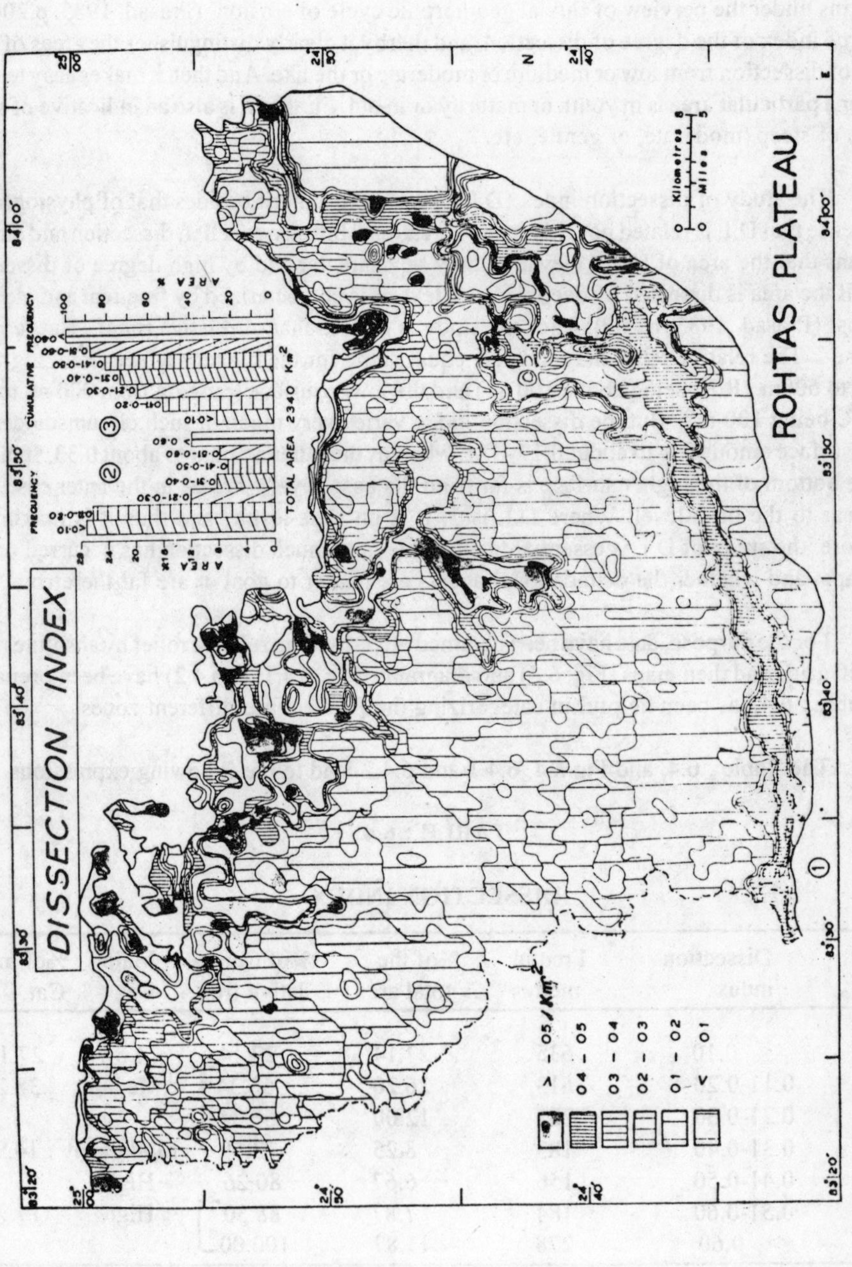
DISSECTION INDEX
ROHTAS PLATEAU
FREQUENCY
CUMULATIVE FREQUENCY
AREA %
TOTAL AREA 2340 Km²
Kilometres
Miles
N
> 05 / Km²
04 – 05
03 – 04
02 – 03
01 – 02
01

FIG. 6.4

1) The dissection index (D.I.) of the Rohtas Plateau ranges from below 0.10 to above 0.60 in general. It ranges more than 0.60 also, but such areas are limited - only 278 grids covering about 278 km^2 (11.87 %). About 65 grids lie in the D.I. zone of more than 0.76.

The Table : 6.4 clarifies that most of the areas (27.14 %) of the Rohtas Plateau lies in low D.I., 38.20 % in medium D.I., 14.92% in moderately high and 19.74% in the zone of high dissection index.

The zone of high I.R. is limited only to 190 grids (one grid of 1 km^2) comprising 5 grids of above 606 m. high, 90 between 551 and 600 m. and 95 grids 501-550 m. I.R. zone.

2) The zone of low D.I. (below $\overline{0}.10$) embraces the areas of southern and eastern alluvial plain. On the Plateau top the low D.I. zone are mostly found on southern and south-western part of the Plateau. These zones are mostly upper reaches of Karamnasha, Durgawati, Sura I and Sura II river basins.

3) The zone of medium D.I. (0.11-0.30) spans over the areas of middle reaches of Karamnasha river basin, upper reaches in Goptha river basin, in the south western part of the Plateau, in Lorsi river basin, in lower reaches of Dhoba river basin **etc.**

4) The zone of Moderately high D.I. (0.31-0.50) embraces the areas of scarps zone, the upper reaches of the Ausane basin, deep notches and **khohs** of all river valleys.

5) The zone of high D.I. (0.51-0.60) spans over the areas of the subsided deep elongated river valleys. The Ausane, Durgawati, Sura I, Sura II, Karsotha, Kohira river basins outside the scarps are subsided river valleys which have have originated due to the collapse of the caverns roofs.

Concluding all it may be said that the low dissection zones are confined either to the Plateau top particularly in the upper reaches of the streams or to the alluvial tracts in the south and east below the scarps. The Plateau top consists of hard Vindhyan sandstones which are resistant to erosion, particularly to those streams which have just initiated and less precipitous. On the other hand the alluvial tracts are devoid of numerous tributaries and frequent hilly exposures.

The high dissection zones are particularly confined to the areas of scarps and notches. The long elongated subsided valleys have been the most dissected areas. Thus we can say that the high dissection zones of the Rohtas Plateau are not the absolute result of denudational processes related to the surface drainage but the phenomena associated with the karstification and later collapsing of caverns as well as tectonic have been responsible for defining such zones of high dissection.

6.5 RUGGEDNESS INDEX

The ruggedness index, too, indicates the degree of dissection of a region, but in this case drainage has also been taken as an important parameter, because the dissection is caused by streams in general. Chorley (1972, pp. 17-90) has, therefore, devised a formula* which has been followed here. Data have been obtained for 2340 grids and then maps (Fig.6.5) and diagram (Fig. 6.5.1 & 6.5.2) have been prepared. The Table : 6.5 has been helpful in categorizing the Plateau into different zones.

TABLE : 6.5

RUGGEDNESS INDEX

No. of Cat.	Ruggedness Index	Freq.in metres	%of total area	%Cumu-lative freq.	Major Cat.	%of major Cat.
I	0.10	1149	49.10	49.10	Low	49.10
II	0.11-0.20	483	20.64	69.74	Medium	32.81
III	0.21-0.30	285	12.17	81.91		
IV	0.31-0.40	141	6.02	87.93	Moderately	9.95
V	0.41-0.50	92	3.93	91.86	High	
VI	0.51-0.60	73	3.13	95.09	High	8.14
VII	0.60	117	5.01	100.00		
TOTAL		2340	100.00			100.00

The Table : 6.5, and Fig. 6.5 and 6.5.1 lead to the following expressions :

1) The ruggedness index of the Rohtas Plateau ranges from below 0.10 to above 0.76 in general. It ranges more than 0.76, but such areas are very limited only in 63 grids covering about 63 km^2. About 54 grids lie in the ruggedness index zone of more than 0.56 to 0.75.

The Table : 6.5 clarifies that most of the areas (49.10 %) of the Rohtas Plateau lies in low ruggedness index, 32.81% in medium, 9.95 % in moderately high and 8.14 % in the zone of high ruggedness index.

2) The zone of low ruggedness index (below 0.10) embraces the areas of south and eastern alluvial plain, the upper reaches of Karamnasha, Durgawati, Sura I, Sura II and Gurwat river basin.

3) The zone of medium ruggedness index (0.11-0.30) spans over the areas of upper reaches of Sura I and Sura II, upper and middle reaches of Dhoba and upper reaches of Lorsi river basin.

4) The zone of moderately high (0.31-0.50) embraces the areas of mostly scarps zone.

5) The zone of high ruggedness index (0.51-0.60) spans over the areas of deep elongated river valleys of Sura I, Sura II, Durgawati and Ausane river valleys.

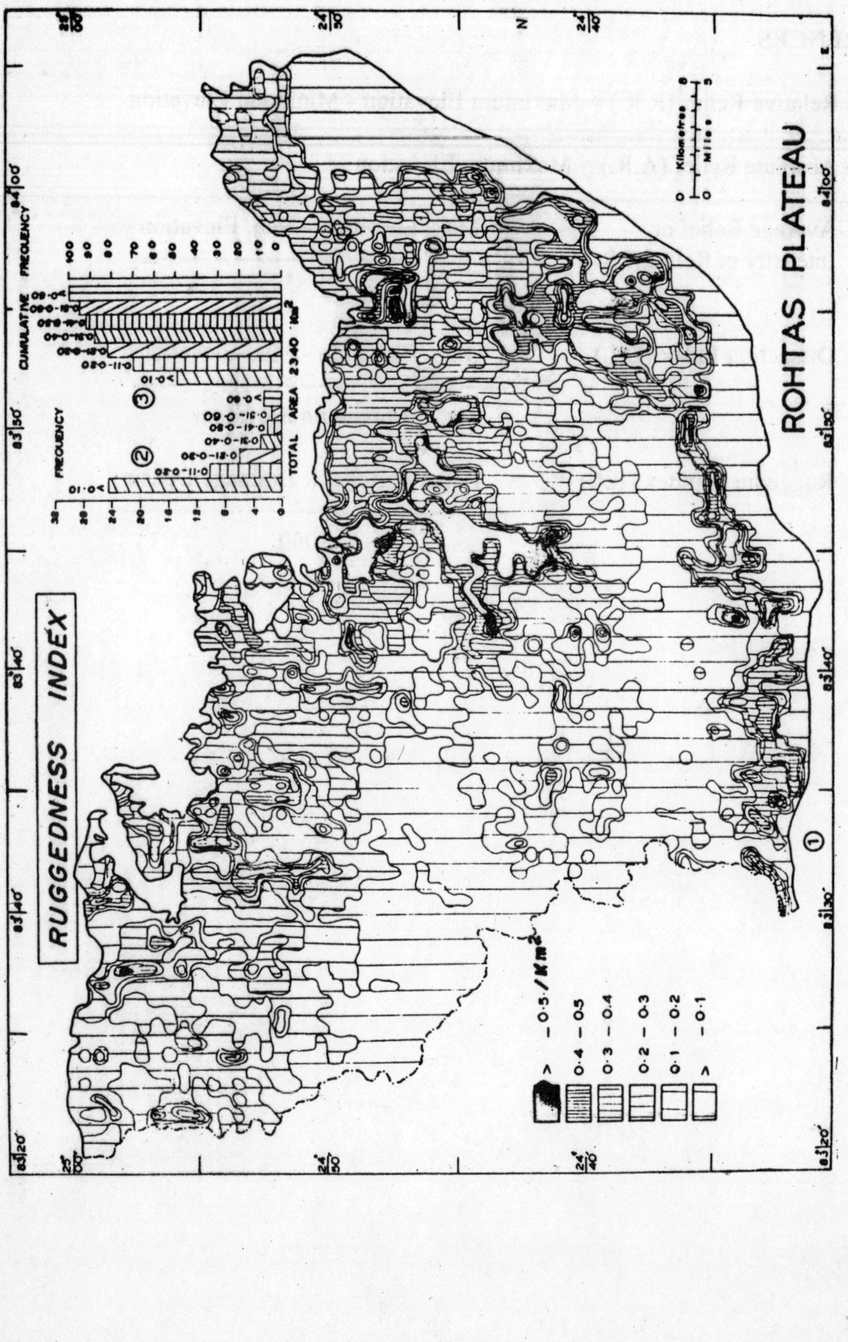
RUGGEDNESS INDEX
ROHTAS PLATEAU
FREQUENCY
CUMULATIVE FREQUENCY

FIG. 6.5

If we observe the maps of dissection index and ruggedness index, the latter is more defined than the former one. The patches of higher ruggedness index have been clearly delimited and these are justified even with the field observations.

REFERENCES

i) Relative Relief (R.R.) = Maximum Elevation - Minimum Elevation

ii) Absolute Relief (A.R.) = Maximum Elevation

iii) Average Relief or Intensity of Relief (I.R.) $= \dfrac{\text{Max. Elevation - Min. Elevation}}{2}$

iv) Dissection Index (D.I.) $= \dfrac{\text{Max. Elevation - Min. Elevation}}{\text{Maximum Elevation}}$

v) Ruggedness Index (R.I.) $= \dfrac{\text{Relative Relief x Drainage Density}}{1000}$

7

SLOPE MORPHOLOGY

'Slopes are ubiquitous elements of the landscape' (King, 1966, p.17), and so is the case with the Rohtas Plateau. Slope steepness influences rates of runoff, soil creep, and soil flowage and the case of cross-country movements of men and vehicles. Nearly all the land surface slopes in one direction or another, and slopes are the fundamental types of landscape feature. The entire landscape is the reflection of slope elements and thus the geomorphology of the Rohtas Plateau is largely concerned with the study of slopes.

7.1 SLOPE FORMS

The slope forms in the study area have been studies by the author by profiling (Figs. 7.1 & 7.2) and contour analysis. During field observation also the character of slope was analysed. The author has drawn profiles at 5' latitudinal and longitudinal intervals by exaggerating the vertical scale 10 times from the normal. An examination of the profile (Figs. 7.1 & 7.2) reveals the following types of slope in the area:

7.1.1. MONOMORPHIC SLOPE

Concave slope

The original scarp slopes might have probably a straight profile top to bottom but denudational processes have modified the slopes in the upper as well as in the lower parts. Since, the limestone and shale beds have weathered rapidly, the lower slope profiles have acquired a concave character. The concavity has developed near the debris slope and continues upward for about 20 metres wherefrom the scarp face stands vertically above.

On the subsided valleys also the hemming slopes make a sharp junction with the valley floor and a clear-cut concavity is visible in the junction area. Concavity is also pronounced when structural controls influence the rate of slope decay. The upper parts composed of hard Kaimur sandstone usually produce straight slope but the slopes start acquiring concave character in the weak shale and limestone beds. The **khohs** described earlier (Ch. 5.2) mainly **Amba khoh,** Jawarh **khoh,** Ropar **khoh** etc. also exhibit concavity of slopes due to change in the character of rocks. Deposition has also helps in imparting concavity in the slope profiles of the **khohs.**

A slight degree of concavity is also visible in the saucer-shaped basins at the top of the Plateau. This concavity, however, is due to tectonic activity.

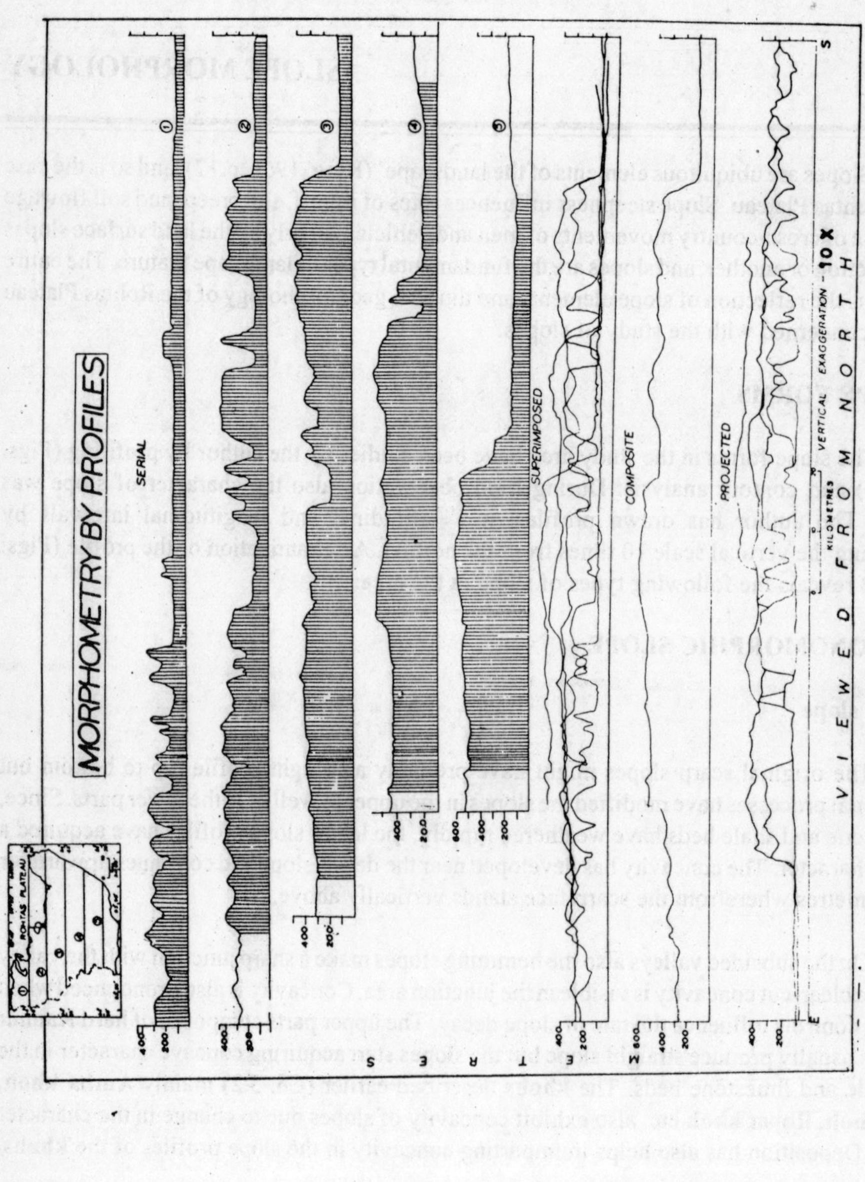
MORPHOMETRY BY PROFILES
SERIAL
SUPERIMPOSED
COMPOSITE
PROJECTED
VIEWED FROM NORTH
VERTICAL EXAGGERATION -10X

FIG. 7.1

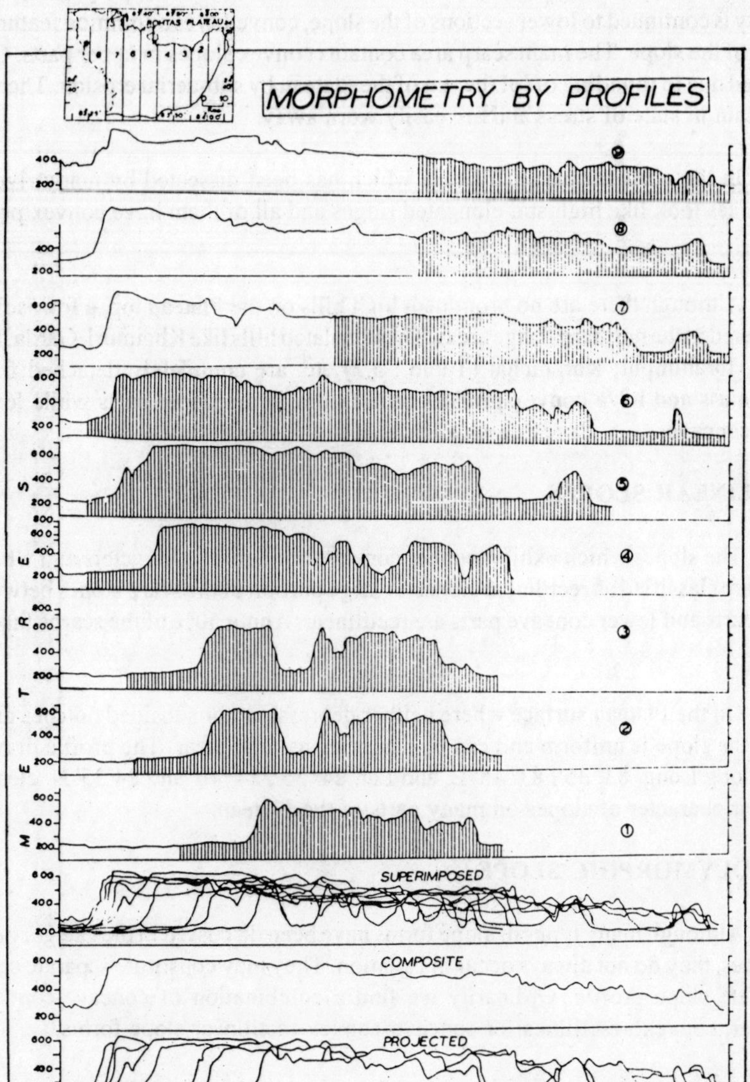
MORPHOMETRY BY PROFILES
METRES
SUPERIMPOSED
COMPOSITE
PROJECTED
S
N
0 10 20
KILOMETRES
VERTICAL EXAGGERATION - 10 X
VIEWED FROM EAST

FIG. 7.2

CONVEX SLOPES

Concavity and convexity are visible in almost all the slope profiles drawn. While concavity is continued to lower sections of the slope, convexity is a common feature in the upper sections of the slope. The main scarp area contains convex slopes in upper parts. Concavity has developed due to rounding off of the rim of the Plateau by sub-aerial erosion. The rock particles also remain in state of stress and are easily worn away.

In the northern part of Plateau which has been dissected by many rivers, the inter-stream areas look like high and elongated ridges and all of them have convex profiles at their top.

Although there are no prominent high hills on the Plateau top, a few isolated hills are encountered in the northern fringe areas. These isolated hills like Khaimuri, Garda, Bhagwanpur, Kalwari, Ibrahimpur, Karamchat (Table : 5.2) etc. are completely detached from the main Plateau mass and have convex profiles on all sides in the upper parts while lower parts are slightly concave.

RECTILINEAR SLOPES

The slopes which exhibit neither concave nor convex characters and which are rather straight are classified as rectilinear slopes. A large portion of the scarp slopes between the upper convex parts and lower concave parts are rectilinear. About 50% of the scarps face has straight slopes.

On the Plateau surface where neither depression nor subsided notches and **khohs** are present, the slope is uniform and can be described as rectilinear. The profile drawn (Figs. 7.1 & 7.2) along Long. 83^0 35', 83^0 45' E. and Lat. 24^0 35', 24^0 40' and $24^0$55' N. clearly show the rectilinear character of slopes on many parts of the Plateau.

7.1.2 POLYMORPHIC SLOPE

Although many types of slope forms have been discussed in the category of monomorphic slopes, they do not always occur in isolation. They may constitute separate components of a complete slope profile. Ordinarily we find a combination of concave-convex, concavo-rectilinear, convexo-rectilinear or concavo-convex-rectilinear slope forms.

Concavo-convex slopes occur widely on the Plateau surface where stream eroded valleys or saucer shaped depressions provide concavity and the adjoining residual ridges provide a convexity in overall slope forms. At the foot of the scarps also especially in the lower half developed on limestone and shale bed the slope is concave in the lower half and convex in the upper. The cobrahood slope discussed earlier is a fine example of concavo-convex slope forms.

Concavo-rectilinear slope type had developed especially along the subsided valleys in the northern half of the Plateau. The slopes along the valleys on the bottom are generally concave. The valley walls are in the form of steep scarps providing the slope rectilinear character. At the junction of the valley floor and valley walls a small ribon-type debris slope has developed which helps assimilate the concave and rectilinear slope componets. Examples of such subsided valley can be seen even in the Ausane and Tutla valley in the south and south-eastern portion also.

The convex-rectilinear slope forms as well developed in the southern scarp face especially in the upper half at the junction of the half resistant sandstones bed at the top and the soft shales in the intermediate area. The slope acquire a convex profile and continues vertically upwards on sandstones. This constitute the rectilinear form of slope belonging to this group. This slope can be categorised as convexo-rectilinear slopes.

If we take the slopes of the scarps from crest to the foot we get three slope components merging with each other. The lower concave and the middle convex slope merge with its rectilinear components continuing up to the top of the scarps. Therefore, we can say that the slope profiles along the high scarps in the south is concavo-convex-rectilinear.

7.2 LINEAR REPRESENTATION OF SLOPE GRADIENT

The correct procedure of representing the gradient of slope is along a line. The linear expression gives exact values. This may be done by drawing profiles either on the basis of contours from maps or by measuring gradient directly in the field with the help of levelling instruments. One such instrument is Abney Level which is handy and suitable for measuring the gradient even of the steep slopes and hill-tops.

The author has drawn a number of profiles (Figs. 7.1 & 7.2) on the basis of contour from the topographical maps (1 : 50000 scale). He also planned to measure the slope with the help of levelling instruments particularly the Abney Level. The instrument was with him also while doing the intensive field work. It is noted that on the top of the Plateau there are no such hills which could be measured for academic benefit. The rest of the Plateau has more or less similar variation in slope. The scarp zones on the other hand are so apprarently steep that the measurement was not needed. Virtualy the profiles drawn on the basis of contours are enough to show the character of gradient of the whole Plateau in general. The profiles (Figs. 7.1 & 7.2) reveal that the gradient ranges from below 2 (minimum 1) in the alluvial tracts and some patches of the Plateau top in the west to 90^0 or so in the scarp zones and walls of subsided valleys. In between these are different segments of slope gradient. Figs. 7.1 & 7.2 present a clear picture.

7.3 AREAL REPRESENTATION OF SLOPE GRADIENT

Although the linear representation of slope gradient gives the exactness of the surface, it does not provide an expression of the space as a whole to which a geographer is attached. We are more concerned with the space and spatial analysis through times is our task. Many trials have, therefore, been taken to provide a spatial analysis of slope gradient. Among the important persons who have shouldered this task are Wentworth (1930, pp. 184-94), Smith (1935, pp. 272-

84), Raisz and Henry (1937, pp. 467-72), Robinson (1948), Strahler (1956, pp. 571-96) etc.

It is, however, noted here that no attempt for the spatial analysis of slope will be absolutely correct in any case. The moments one will deviate from a line and will generalise the gradient of a particular area, the result will be deviated from the truth. It is natural because the slope is so variable even in a shortest line that an areal picture cannot be true having so much lines in different directions. Therefore, the devices which have been presented can be called as 'trials' but at the same time the credit goes to those devices which are nearer to the truth. The devices presented by Wentworth and Raisz and Henry are the most popularly acclaimed devices among all the working methods in geomorphological researches. The author has, therfore, prepared two maps following these two methods, although he is more satisfied with the Raisz & Henry method because of its near-reality with the field observations.

7.3.1 AVERAGE SLOPE

Slope gradient over the entire Plateau area has been shown by means of an *Average slope* map (Wentworth, 1930, pp. 184-194).

This gives distribution of the degree of surface inclination. No doubt, the construction of this slope map is extremely time consuming. The 1 : 50000 Topographical sheets of the area have been divided into two centimetre grids, and the contour-crossing are counted to determine the average number of contour-crossing per kilometre. The result from the formula* gives the average slope in terms of the tangent of the average angle of slope which has been converted into degrees with tangent tables and scientific calculator. This technique seems to be very much useful in making quantitative generalisation.

The prepared map (Fig. 7.3) shows that the average slope in the Rohtas Plateau ranges from below 2^0 to above 12^0. The highest gradient observed is in the scarps area, then in the dissected notches and khohs of the Plateau. The lowest gradient is in the lower portion of the alluvial tracts.

The Table : 7.1 shows the percentage of the total area having different degrees of slopes. About 9.74 % of the total area has a slope of below 2^0. Similarly 14.24 % of the total area has a slope 2.1 - 4.0^0 ; 17.27 % has 4.1 - 6.0^0 % ; 17.22 % has 6.1 - 8.0^0 % ; 16.36 % has 8.1 - 10.0^0 % , 9.02 %has 10.1 -12.0^0 and 16.15 % of the total area has more than 12^0.

Similarly Table : 7.1 shows the four categories of Average slope zones, viz., steep, moderately steep, gentle and moderately gentle covering 0.95, 24.22, 50.85 and 23.98 per cent respectively to the total area of the Rohtas Plateau.

---

* Wentworth devised a formula :

$$\text{Average Slope} = \frac{\text{Average number of contour crossing / km. X contour interval}}{636.6}$$

TABLE : 7.1

AVERAGE SLOPE
(Wentworth)

Slope cat. in degree	Freq. in kms²	%of the total area	% Cumulative freq.	Major slope cat.	% of the major slope cat.	Cumulative freq. %
2 & below	228	9.74	9.74	Moderately	23.98	23.98
2.1 - 4	333	14.24	23.98	Gentle		
4.1 - 6	404	17.27	41.25			
6.1 - 8	403	17.22	58.47	Gentle	50.25	74.33
8.1 - 10	383	16.36	74.83			
10.1 - 12	211	9.02	83.85			
12.1 - 14	148	6.32	90.17			
14.1 - 16	101	4.31	94.48	Moderately	24.22	99.05
16.1 - 18	59	2.52	97.00	Steep		
18.1 - 20	48	2.05	99.05			
> 20	22	0.95	100.00	Steep	00.95	100.00
TOTAL	2340	100.00			100.00	

Analysing both the Table and map (Table : 7.1 & Fig. 7.3), the following observations may be made :

i) The steep slope category above 20^0 covers an area of 22 square kilometres (0.95 %). The steep slope is mostly found in the southern, south-eastern and northern scarp zones, the walls of all the notches and **khohs** of the river valleys like upper reaches of Goptha, Ausane, Karsotha, Sura I, Sura II and Dhoba rivers.

ii) The moderately steep slope zone (10.1^0 - 20.0^0) covers an area of 567 square kilometres (24.22 %). This zone is mostly found in the upper reaches of Durgawati, Karamnasha, Koel and Gurwat river basins. It is also found in the lower portion of steep slope, lower portion of the walls of notches and **khohs.**

iii) The gentle slope zone (4.1 - 10.00) covers an area of 1190 square kilometres (50.85 %) and is found in the upper portion of debris slope, lower subsided valleys of all river basins, top of the Plateau in the depressed saucer-shaped areas, upper portion of Karamnasha, Dhoba, Koel, and other river basins.

iv) The moderately gentle slope zone (below 4^0) covers an area of 561 square kilometres (23.98%). This zone is found in the lower portions of debris or talus slope and in the whole alluvial tracts.

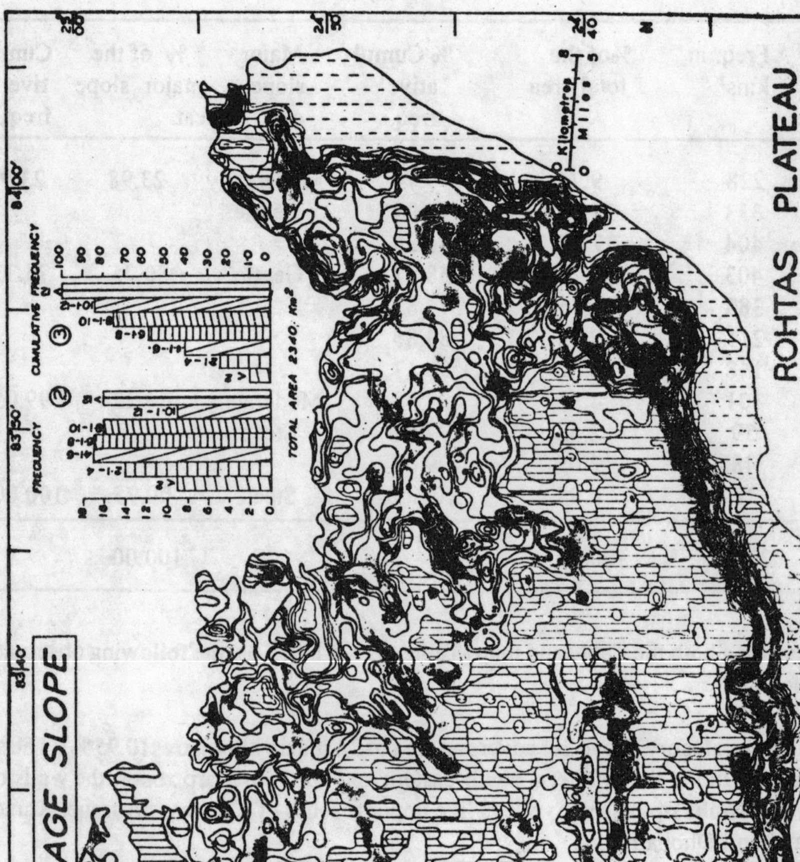
AVERAGE SLOPE
ROHTAS PLATEAU
SLOPE IN DEGREES
> 6
4 – 6
2 – 4
< 2
Wentworths Method
FREQUENCY
CUMULATIVE FREQUENCY
TOTAL AREA 2340 Km²
Kilometres
Miles
N

FIG. 7.3

The Wentworth's method of slope analysis thus provides only four major categories having 24° as maximum. This is far from the truth when one bothers to visit the area. The author has mentioned earlier many times, on the basis of his field observations, that there is a considerable percentage of area characterised with bold and vertical escarpments and valley walls. This nature of slope has been eliminated by the Wentworth's method. This is a major flaw of this method. Therefore, the author has given minimum weightage to the findings obtained from this method while discussing the evolution of slope and other related aspects.

7.3.1 SLOPE CATEGORY

Following the Raisz and Henry method, the author divide the large scale topographical maps into small regions, within each of which the contour lines had some standard spacing i.e., the same number of contour lines had some standard spacing per kilometre of horizontal equivalent.

Though this method requires a thorough identification of the contour density zones, the procedure of construction is rather simple. At first, a horizontal scale of standard contour spacing is drawn. The next step is associated with the zonation of each slope category according to scale of standard contour spacing with the help of divider (Fig. 7.4).

Accordingly seven major categories have been obtained (Table : 7.2). The Table 7.2 shows the percentage of the total area having different degrees of slopes. About 20.94 % of the total area has a slope of below 2°. Similarly 21.39 % of the total area has a slope of 2.1 - 4.0° ; 12.75 % has 4.1 - 6.0° ; 13.07 % has 6.1 - 8.0° % ; 5.26 % has 8.1 - 10.0° ; 10.80 % has 10.1 - 12.0° ; 4.7 % has 12.1 - 20.0° ; 6.54 % has 20.1 - 45.0 and 4.50 % of the total area has more than 45°.

TABLE : 7.2
SLOPE CATEGORY
(Raisz & Henry)

Slope cat. in degree	Freq. in kms2	% of the total area	% Cumulative freq.	Major slope cat.	% of major cat.	Cumulative freq.,%
2 & below	489.99	20.94	20.94	Moderately	42.33	42.33
2.1 - 4.0	500.59	21.39	42.33	Gentle		
4.1 - 6.0	298.15	12.75	55.08			
6.1 - 8.0	305.97	13.07	68.15	Gentle	31.08	73.41
8.1 -10.0	123.18	5.26	73.41			
10.1 -12.0	252.75	10.81	84.22	Moderately	15.55	89.96
12.1 -20.0	110.70	4.74	88.96	Steep		
20.1 -45.0	153.17	6.54	95.50	Steep	6.54	95.50
> 45.0	105.50	4.50	100.00	Very Steep, Vertical & Cobra-hood	4.50	100.00
TOTAL	2340	100.00			100.00	

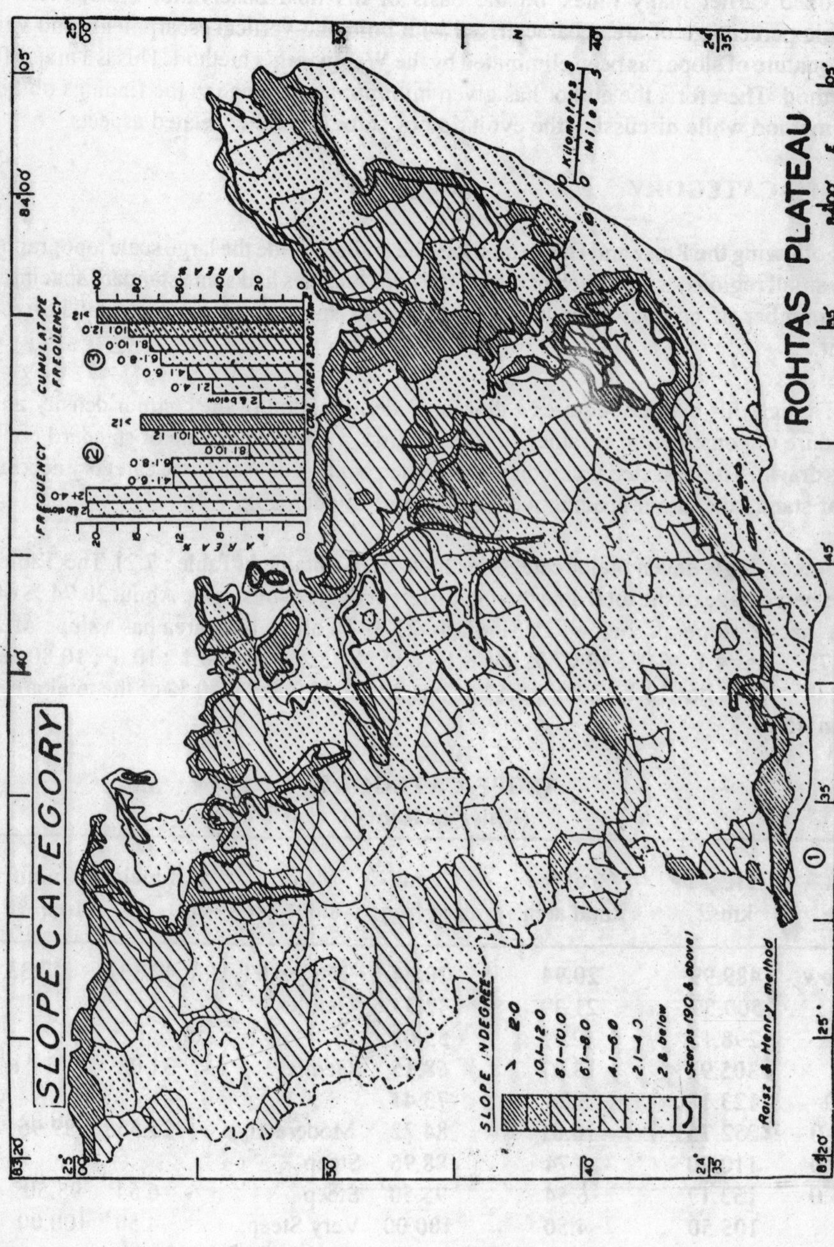
SLOPE CATEGORY
ROHTAS PLATEAU
SLOPE IN DEGREE
Scarps (90 & above)
Raisz & Henry method
FREQUENCY
CUMULATIVE FREQUENCY
AREA %
Kilometres
Miles

FIG. 7.4

Similarly Table 7.2 shows the seven categories of slope, **viz.**, cobra-hood, vertical and very steep covering 4.50 % together and steep, moderately steep, gentle and moderately gentle covering 6.54, 15.55, 31.08 and 42.33 per cent respectively to the total area of the Rohtas Plateau.

Analysing both the Table and map (Table : 7.1 & Fig. 7.4), the following observations may be made :

i) Although the slope category of cobra-hood, vertical and very steep gradient comprises only 105.50 square kilometres (4.50 %), it is the most fascinating and perceptible zone in the study area. This is mainly confined to the **escarpments and** notches, particularly in those areas where there are outcrops of limestone and **shale beds under** the Vindhyan sandstone capping. This character has resulted in maintaining **the outward** projection of sandstone capping over the backward eroded limestone and shale. **The** gradient is very steep where the foot of the escarpments is covered with debris; it becomes vertical where the debris is washed away from the foot and where there is case of slumping sandstones; and the slope looks like cobra-hood where there exists outward projection of the sandstone capping. The cobra-hood slope is found mostly in the areas of southern lower scarp face.

ii) The gradient of steep slope ranges from 20.1 and 45.0^0 and covers an area of 153.17 square kilometres (6.54 %). The steep slope zone is mostly found in the patches of south western scarp zone, east of Bhawatia river on the scarp face, southern scarp face, east of Rohtasgarh scarp face, upper reaches of Ausane river, east of Banjari hill, Scarp face in the west of dumarkhar river, northern scarp face, the wall of nothcs & **khohs** of Durgawati, Koel, Karsotha, Sura I, Sura II rivers.

iii) The moderately step slope zone (10.1^0 - 20.0^0) covers an area of 363.45 square kilimetres (15.55 %). This zone is mostly found in south western, southern and eastern scarp face in the extreme north eastern scarp face, east of Dumarkhar river on the scarp face, in isolated hill slopes, and in the north-western scarp face. It is also found on the Plateau top in small patches such as in the upper reaches of Dumarkhar river, in the northern tributaries of Ausane river and on the southern margin of Sital kund.

iv) The gentle slope zone (4.1 - 10.0^0) comprises an area of 727.30 square kilometres (31.08 %), and is found in the upper portion of debris slope, lower portion of subsided valleys of all the river basins. On the Plateau top this category is found everywhere.

v) The moderately gentle slope zone (below 4^0) includes an area of 990.53 square kilometres (42.35 %), and is found in some patches of the Plateau top in the west, in the alluvial tracts, in the lower reaches of Durgawati valley and Sura I outside the scarp zone.

The preceding discussion on the slope gradient concludes that the Rohtas Plateau is such a region where slope of various inclinations ranging from the monotonous alluvial tracts to the vertical escarpments is observed. It can also be concluded that the method devised by Raisz and Henry is more suitable and accetable than that of Wentworth for the areal analysis of slope on the basis of contours from maps.

7.4 EVOLUTION OF SLOPE

'Evolution concerns the change of slope form with time. One of the principal questions is whether a slope, under given circumstances, becomes steeper, less steep, or retains the same form after it has been acted upon by processes for a period of time' (Young, 1972, p.25). Rohtas Plateau was uplifted as a flat tablenland in the Tertiary period with a slight inclination towards the north. During the process of uplift the original scarps were produced and there was perhaps no debris slope in the Tertiary period. Originally, therefore, the Plateau contained two kinds of slopes - one along the tectonically produced scarps and the other a general slope towards the north. There is no evidence to suggest that the slopes associated with subsided valleys, *khohs*, notches and saucer shaped basins existed before the Tertiary period. Due to the special geological structure of the area and succession of geological beds in particular order, the underground caves, however, might have formed before the Tertiary earth-movements took place. During the violent shocks generated by the Tertiary orogeny, the Plateau area gained further height but the overlying sandstone strata also subsided over underground caverns, already in existence. Since then the types of slope increased.

'Slope is a function of structure, process, time and tectonics' (Ahmad, 1985, p.41). All these factors have played their roles in the formation and characteristics of slopes found in the Rohtas Plateau. Structure has exercised a dominant control on the evolution of slopes because of the presence of underlying limestone strata which are soluble. The solution and draining out of limestone was responsible for formation of caves and subsidence of overlying rock materials giving birth the Tertiary. Structure has also introduced modifications in the tectonically produced scarps by changeing its straight face into a concave, cobra-hood, rectilinear-convex face. This was caused by the rapid erosion in the weak shales and limestone below which differ from the strong sandstones above.

The agents which bring changes in the form of slopes are discussed as processes of slope evolution. They include all thsoe activities, mechanical and chemical, which slowly modify slope charaters. Solution, oxidation, soil creep, soil wash, rock-fall, rock-slides are some of the processes which are responsible for changes in slope morphology. The Rohtas Plateau is situated in the tropical humid climatic region where fluvial action is the most dominant factor in shaping the form of slopes. It is due to solution of limestone that underground caves were formed and large scale subsidence took place in several parts of the Plateau. Had there been an arid climate solution would not have been possilbe and the Plateau surface would have exhibited a flat surface. Saucer-shaped depressions, depressed valleys,notches and **khohs**, all these features have been produced by water action.

The rapid wearing down of the lower scarp face is also due to the fact that water has been able to effect weak shale and limestone beds quickly. Water action has thus modified the scarp slope in the lower part.

Time element is also an important aspect which affects slope forms. As has been mentioned earlier the debris slope at the foot of the scarps was not in existence before the Tertiary upheaval. After the formation of scarps the debris slope has been gradually built up by accumulation of talus and scree at the foot of the scarps. This must have taken a sufficiently long time. The area and extent of the debris slope may get further modified in future because the amount of debris will continously increase. Time has also been responsible for carving out of the river valleys on the Plateau surface. At the time of uplift during Tertiary the upper portions was almost featureless. With the passage of time, however, the area is becoming more and more picuresque by creation of valleys, **khohs,** notches, basins **etc.** In very distant future the lower limestones may be completly consumed by water and it is very difficult to predict the character of slopes which may be formed by such a happening.

Tectonic activity has shaped the initial slopes of the Plateau. The bold and prominent vertical slope in the scarp area is an example. This section is the product of tectonic activity. The slopes developed in depressed valleys and saucer-shaped basins also owe their origin to tectonics.

While one examines the causes responsible for the slope evolution, he has also a task to see whether the existing slope elements are the product of porcesses different from the present ones. Many theories have been propounded to help in understanding the past. Important among them are those presented by W.M. Davis (1892, 1898), W. Perick (1924) and L.C. King (1957). Davis contended that evolution of landscape takes place by the process of slope flattering which changes a V shaped cross profile of a river into shallow curves by water action. He did not present a systematic theory of slope development. His ideas are contained in his general theory of landscape evolution.

When we examine the Davis ideas in relation to the slope features of the Rohtas Plateau we find that the scheme of slope flattening envisaged by him is not visible in the area. Davis emphasised the role of fluvial action in shaping the main features of landforms. Fluvial action is not acting freely in the Plateau area because water derived from rainfall percolates down through the joints and cracks of the overlying sandstone strata and initiates solution process in the limestone strata below. Geological work of flowing water on the Plateau surface is limited. The main features of slopes were produced by uplift along the scarps downwarping on the Plateau surface and collapse of cavern roofs downwards. These features are not a product of water action in the sense Davis described them. A few valleys have been carved out on the Plateau surface which have not attained depth so far. These valleys have a V shaped cross profile on the plateau. The V shaped shall remain maintained till the depth of valley does not increase and till the time the valley depth remains confined within the strong sandstone beds. The process of slope flattening may be taking place in the sandstone area. As soon as the valley is able to cut through the upper sandstone and reaches the lower limestone area the sequence of developments of slopes will change drastically. This will be so becasue the normal fluvial process of water erosion will be taken over by processes of solution and draining out of limestone. It is probable that slope flattening process will stop and gorge-like features will be produced as we find in the northern reaches of some river valleys.

Penck introduced the idea of slope replacement. In this scheme the initial slope has replaced by another slope after complete decay of the former slope. His theory is based on the degree of reduction of regolith, the mobility of regolith, the regolith thickness, the exposure of the slope surface to weathering processes, the character of rock, the type of climate and the angle of slopes. 'Penck assumed that as soon as rock waste resting on a slope of given angle attains the mobility necessay to be affected by denudation it is instantaneously removed from the slope' (Young, 1972, p.31).

Since the development of slopes in the study area has taken place mainly due to tectonic activity the scheme of slope replacement envisaged by W. Penck does not seem to apply here. Replacement of slope is neither visible in subsided caves nor in saucer shaped basin nor notches and *khohs*. They may be in operation on the scarp face but L.C. King's theory of parallel retreat of slopes is better able to described the processes operating in those areas.

L.C. King's concept of parallel retreat of slopes was based on the works of wood (1942), and Fair (1947). Fair had examined the scarps of South Africa and found them retreating in a parallel plane. King applied this idea to formulate a general theory of slope development and stro gly advocated the idea of parallel retreat of slopes.

Kings's ideas may be applied in the study area and they may be examined with reference to the behaviour of the scarp face since its origin in the Tertiary period. Accumulation of huge quantity of debris at foot of the scarps suggests that the materials have been derived from the wear and tear of the scarps face to produce a wide area of debris slope, huge quantity of talus and scree must have been needed and these materials have perhaps been derived by the weathering of scarp face. Since the main scarps exhibit a 90^0 slope even today, this can be safely assumed that the scarp is retreating parallel to itself. The preservation of the initial slope since the Tertiary period explains the idea enuciated by king. The scarps are perhaps retreated in the same way as the scarps of the Drakenberg are retreating in South Africa. There are evidences to suggest that the scarps have retreated considerably in southern part away from the Son river.

On the Plateau surface also the depressed valley slopes are retreating because the gorge-like structure is progressively becoming wider.

Attempts have been made by several writers to identify the elements of slope. According to the definition of savigear (1960), 'an idealized model of a fully developed normal slope reveals three distinct sections of slopes ; (i) Crest slope extending from interfluves to the point where the slopes steepens (ii) the back slope consisting of steepest units of the profile and (iii) the foot slope, a gentle inclined section extending from the back slope, (Prasad, 1979, p.244) Wood (1942), however has recognised four elements in a slope - the upper convex portion called waxing slope, the steep central portion called free-face, the lower debris slope called debris or talus slope and the outer gently sloping area called waning slope or wash slope.

The slope elements recognised by L. Wood can be traced in the scarp zone of the Rohtas Plateau. The upper convex portion of the slope along the rim of the plateau can be conveniently termed waxing slope. The vertical sections of slope below the convex slope composed of hard and bare rocks free from vegetation cover can be taken as the free-face of the slope. In the lower portion of the free-face slight concavity has developed due to differing geological structures along the slope face. There is a wide ribbon of debris slope in the southern parts which has been produced by the accumulation of debris derived from the waxing slope and the free-face. The last element of the slope which is gentle and separates the scarps from the beds of the Son river can be described as waning slope.

Thus the slope elements are not a product of erosion alone. To a large extent the elements have been introduced by tectonic activity. Erosional processes have added some elements to the original slope.

8

GEOMORPHIC REGIONS

We have examined, in detail, the geological evolution, surface elevation, geology and structure, tectonic history, topography and morphometry (**i.e.,** slope, profiles and drainage **etc.**) of the Rohtas Plateau. These are the fundamental bases upon which the area may be divided into geomorphological regions.

A geomorphic region may be described as a part of the earth's surface with recognisable geomorphological qualities which distinguish it from adjacent regions. Similar to and rather in different way from this definition, it may be termed as a region of homogeneous geomorphic unity, **i.e.**, unity in almost all aspects of geomorphology such as geology, structure, relief, hydrology and morphometry, **etc.** and to some extent climate also, the role of which cannot be ignored. 'It is the combined existence of these features that determines regional differentiation or the individuality of the regions in geomorphology also' (Ahmad, 1965, p.309).

No doubt, the Rohtas Plateau itself is a geomorphic region in the highlands of Bihar. But from the foregoing chapters it is apparent that within the delineation of the Plateau there are variations in materials, forms and processes. The Plateau top has capping of hard old red Vindhyan sandstone whereas the limestone-shale beds below are peeping through the scarps. On the foot are the debris slopes which are fringed by new alluvial tracts. Both endogenetic and exogenetic processes have evolved a variety of forms ranging from a near-flat alluvial zone to the high vertical rocky escarpments. In between these are the highly and moderately high dissected plateaus, subsided valleys, notches and **khohs,** resistant outcrops in the form of ridges and hills, karst depressions and many other large and small details of topographic features. The morphometric variables like drainage texture, hydrographic network, relative relief, dissection index, ruggedness index, average slope and many other aspects of basin morphometry clearly depict the spatial variations within the Plateau. Both erosional and depositional activities have been noted. All these variations in materials, forms and processes have been depicted on the Geomorphological Map (Fig. 8.1). It is noted that this geomorphic map (Fairbridge, 1968, pp. 388-403) has been prepared using the standard symbols following the recommendation of the international Geographical Unions's Subcommission on Geomorphological Mapping which was set up at the 1960 Stockholm meeting. This map is virtually presenting a broad outline of geomorphic regions in the Rohtas Plateau. But many aspects of geomorphic significance are missing, rather all the minute details cannot be represented in a map. Therefore, the author thought it better to divide the study area into its geomorphic regions taking all the details into considerations. It was rather a tough task for him, but he had numerous tools, particularly drawn maps and field reports with him. He could superimpose the maps over one another to

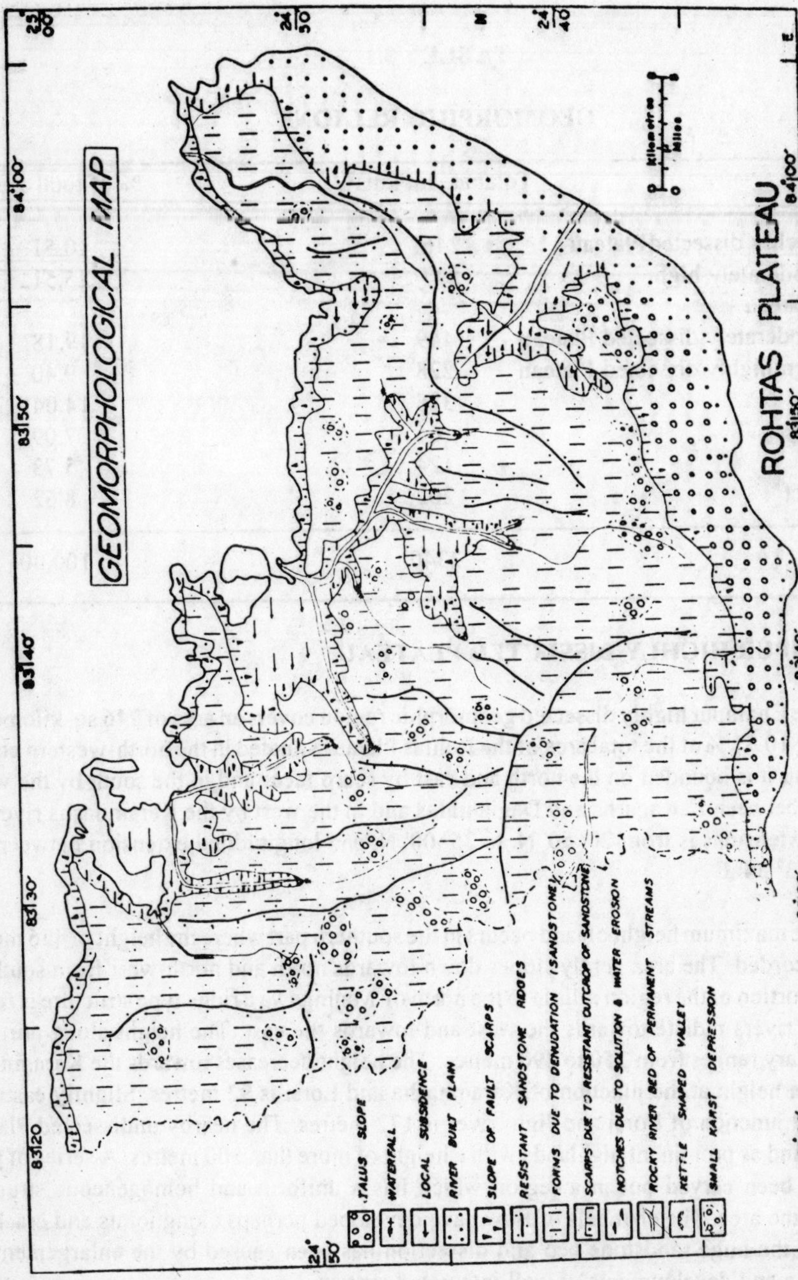
GEOMORPHOLOGICAL MAP
ROHTAS PLATEAU
TALUS SLOPE
WATERFALL
LOCAL SUBSIDENCE
RIVER BUILT PLAIN
SLOPE OF FAULT SCARPS
RESISTANT SANDSTONE RIDGES
FORMS DUE TO DENUDATION (SANDSTONE)
SCARPS DUE TO RESISTANT OUTCROP OF SANDSTONE
NOTCHES DUE TO SOLUTION & WATER EROSON
ROCKY RIVER BED OF PERMANENT STREAMS
KETTLE SHAPED VALLEY
SMALL KARST DEPRESSON
83°20'
83°30'
83°40'
83°50'
84°00'
24°30'
24°40'
24°50'
25°00'
N
E
Kilometres
Miles
0

FIG. 8.1

have an approximate common boundary. He could verify and alter it on the basis of field reports. In this way, the author could delineate eight major geomorphic regions in the Rohtas Plateau. (Fig. 8.2).

TABLE : 8.1

GEOMORPHIC REGIONS

Regions	Total areain kms²	% of total area
Chainpur highly dissectedPlateau	246	10.51
Adhaura moderately high dissected Plateau	597	25.51
Southern moderately dissected Plateau	449	19.18
North-eastern highly dissected Plateau	228	9.40
Scarp zone	328	14.04
Subsided valley	166	7.09
Debris slope	124	5.73
Alluvial tract	202	8.62
TOTAL	2340	100.00

8.1 CHAINPUR HIGHLY DISSECTED PLATEAU

The Chainpur highly dissected geomorphic region covers an area of 246 sq. kilometres representing 10.51 % of the total area of the Rohtas Platau. Situated in the north-western corner of the Plateau it is bounded on the north and east by scarp faces and in the south by the water parting line between Lawagarha and Dagia **nalas** and in the west by the Karamnasha river. Its latitudinal extensions is from 20° 50' N. to 25° 00' N. and longitudinal extension between 83° 20' E. and 83° 34' E.

The maximum height of land occurs in the southern part where the height of 436 metres has been recorded. The area gently slopes down towards north and north-west from south. In the central portion of the region a little to the north of Majhigawa a ridge type structure is found from where rivers radiate towards the west and towards the east. The height along northern scarp boundary ranges from 250 to 290 metres. The height decreases towards the Karamnasha river and the height at the junction of Karamnasha and Lorsi is 92 metres. Slightly eastward the height at junction of Lorsi and Jigni river is 172 metres. The nearby undissected Plateau remnants stand as prominent highlands with a height of more than 300 metres. A series of river valleys has been carved out in a region which has a uniform and homogeneous structure throughout the area. The channels of flow have developed perhaps along joints and cracks in the body of the huge sandstone bed and dissection has been caused by the enlargements of drainage line and developments of well integrated system.

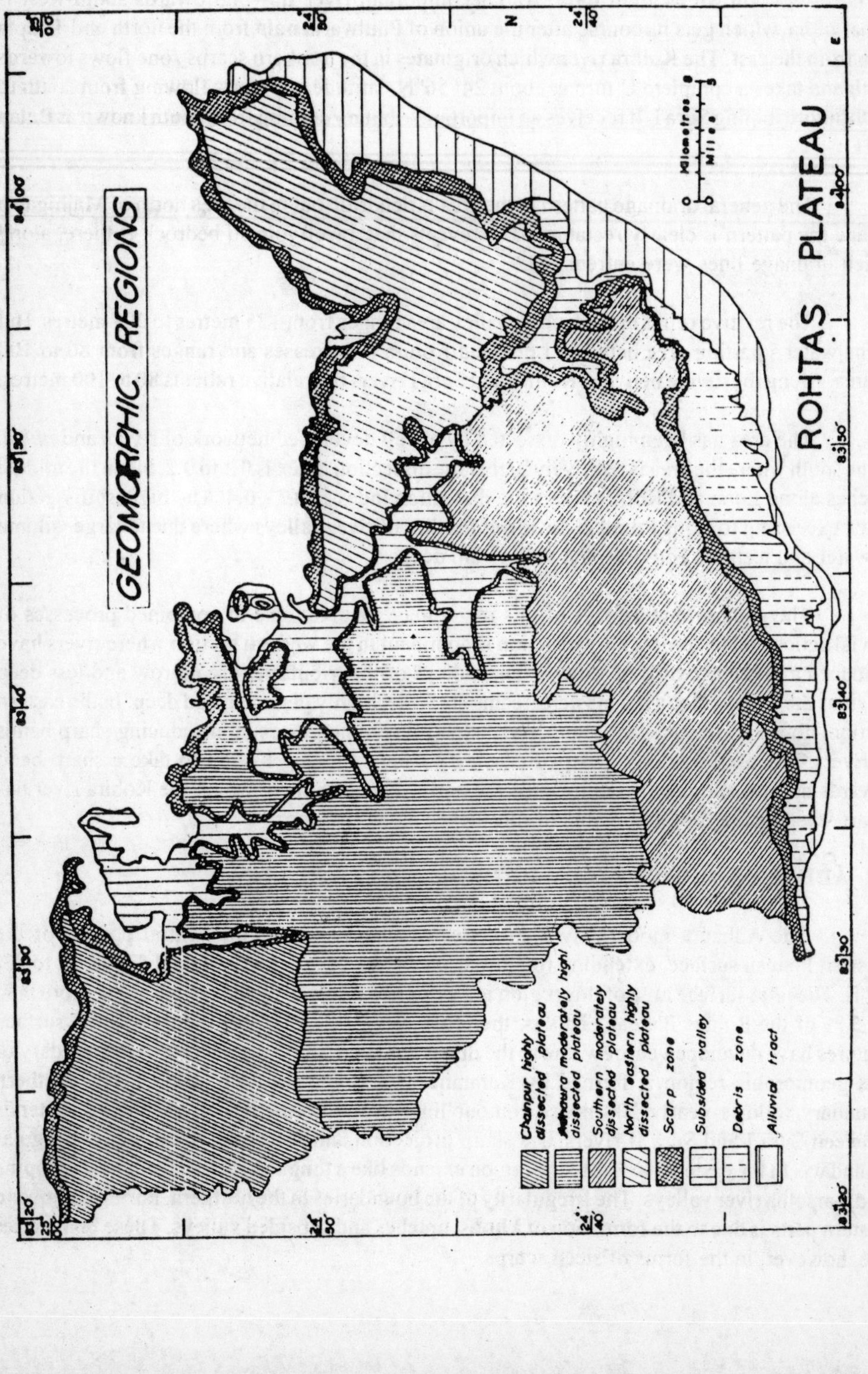
GEOMORPHIC REGIONS
ROHTAS PLATEAU
Chainpur highly dissected plateau
Adhaura moderately high dissected plateau
Southern moderately dissected plateau
North eastern highly dissected plateau
Scarp zone
Subsided valley
Debris zone
Alluvial tract
Kilometres
Miles
N
E

FIG. 8.2

The main river flowing to the west is Lorsi which receives an important tributary coming from the north-east known as Jigni **nala.** Another important river flowing towards south-west is Okhargarha which gets its course after the union of Phulwaria **nala** from the north and Thupta **nala** from the east. The Kohira river which originates in the northern scarps zone flows towards south and takes a complete U turn at about 24^0 56' N. latitude and starts flowing from south to north before joining Sura I. It receives an important tributary coming from south known as Belan river.

The general drainage pattern in the area is dendritic but in the area north of Majhigawa village the pattern is clearly rectangular. It reveals that a well jointed bedrock is there, along which drainage lines were entrenched.

The relative relief in the Kohira valley area ranges from 125 metres to 203 metres. But in the water - parting area between Lorsi and Kohira it decreases and ranges from 80 to 100 metres. Along the divide between Kohira and Sura I rivers the relative relief is 80 to 100 metres.

The area has been highly dissected by a well developed network of rivers and *nalas*. In the south where the area is relatively higher the dissection index is 0.1 to 0.2, but in the middle reaches along Lorsi and Kohira rivers the dissection index is 0.3 - 0.4. The highest dissection index exceeding to 0.5 is found in the lower reaches of river valleys where due to large volume of water and tectonic activities the valleys are deeper.

The geomorphic aspects of this region have evolved due to combined processes of fluvial action and tectonics. Fluvial action is dominant in the western section where rivers have produced a dense network of valleys. The valleys of the tributaries are narrow and less deep whereas the valleys along the master streams are wide, highly dissected and deep. In the eastern section, however, tectonic activities have helped fluvial processes in introducing sharp bends in rivers. The area was downwarped causing a south flowing Kohira to take a sharp bend towards the east and then again towards the north. The area along which the Kohira river has a east-west course is porbably due to subsidence along the line.

8.2 ADHAURA MODERATELY HIGH DISSECTED PLATEAU

The Adhaura moderately high dissected Plateau covers the central portion of the western Plateau surface, extending roughly from 24^0 37' N. to 24^0 55' N. and 83^0 23' E. to 83^0 41' E. The total surface area of this region is 597 square kilometres and covers about more than 25.5 % of the Rohtas Plateau. This is, therefore, largest geomorphic unit where the surface features have developed entirely due to the dissection work of rivers. The western boundary of this geomorphic region is formed by Karamnasha and Gurwat river valleys. The southern boundary follows nearly 500 metre contour line. In the northern side the Plateau extends between Sura I and Sura II rivers as a sharp projection and therefore, has a highly irregular boundary. In the eastern parts also the region extends like a tongue between Durgawati, Goptha and Karsotha river valleys. The irregularity of the boundaries in the northern, north-eastern and eastern parts is due to the formation of **khohs,** notches and subsided valleys. These boundaries are, however, in the forms of steep scarps.

The surface evelvation varies moderately throughout the Plateau. At the head of Sura I the height is 444 metres. It acts as a water-parting zone between Sura and Karamnasha basins. The height decreases towards south of this area and increases up to 495 metres in the area of the ridges which extend in south-west-northeast direction. South of this highest ridge in the west central portion of the region the tributaries of the Karamnasha have been able to dissect the Plateau to some extent and the height decreases to 380 metres along the course of the Karamnasha rivers. It starts increasing again southwards and along the southern boundary of the region it rises to 511 metres.

Between Sura I, Sura II rivers the height of the projected scarps is 338 metres. Going straight to southward regions the height goes on increasing because the entire zone from north to south acts as a waterparting area between Karamnasha systems flowing to the west and Karsotha, Koel, Hasadah, Hathidah to the north-east. The average height of this water parting zone is 510 metres and the highest point is observed along the southern boundary where it is 543 metres. A little to the west of this waterparting zone is the Adhaura depressions where the height is 459 metres. From east to west there are other depressions like Lohra (490 m.). Chainpura (400 m.) and Harbag (390 m.). since the difference of height between the highest point that is 540 metres and the lowest portion **i.e.** 390 metres is about 153 metres, we classify this geomorphic region as a moderately high dissected region. Although the rivers have not been able to carved deep valleys the drainage system is well entrenched and dissection is taking place.

Along the eastern section the height is 443 metres near Koel river which increases to 510 metres along the ridges and decreases again along Khora river. The difference of height of this region is only 68 metres.

Most of the river like Karamnasha, Khora and Koel exhibit dendritic drainage pattern. In the Sura I we find herring bone pattern which is due to structural control. An area of radial drainage can also be spotted in the region from where Sura I, Sura II, Koel, Karamnasha and tributaries of above noted rivers as well as the tributary of Khora river radiate on all sides.

In the northern portion of the geomorphic region the vegetation is of dense mixed variety where different species of plants are found. In the area north of Karamnasha river the forests is of dense mixed (mainly **sal**) while an area of open mixed forest lies between these two vegetation belts. South of the **sal** belt also the forests are of open mixed variety and extend up to the southern boundary of the region.

The relative relief varies slightly from area to area. In the Sura I, Sura II and Karsotha river basins the relative relief ranges between 180 m. and 240 m., while in the central portions of the region where difference of height in small the relative relief is 20 to 40 metres. In the highly dissected north-eastern part as well as along the valleys of Gurwat the relative relief is 60 to 80 metres. The lower reaches of Karamnasha it is 80 to 100 metres.

The dissection index for the geomorphic region as a whole averages from 0.1 to 0.3. But there are areal differences. In the north-western portion west of Sura I basin the index is 0.4. In the upper reaches of Sura I, Sura II, Karsotha and Gurwat rivers the index is 0.4 to 0.5.

In the northern and north-eastern scarps zone the index is 0.5. This means that the higher elevation areas show high degree of dissection.

The area has been evolved and shaped by fluvial action. The upper surface which is composed of hard sandstones has yielded to the channelised flow of water. This has carved the region into valleys and ridges. The inter-stream areas which withstood the attack of erosive forces are visible on the surface in the form of elongated ridges with a relatively higher elevation. Tectonic activity has added the evolution of the region of the northern, north-eastern and eastern boundaries where subsidence has produced notches, **khohs.** The irregular boundary of this geomorphic region towards the north, north-east and east is due to tectonic activity. The tectonic activity, however, is not the only controlling factor in the evolution of the geomorphic features of this part. The fluvial activity specially its activity on limestone strata is the dominant factor. The caves were produced by the solution of limestone which due to tectonic activity of Tertiary period subsided to give an irregular shape to the northern and eastern boundary of this geomorphic region. Such large scale tectonic effect is not visible in the western part of the region, but in several places, depressions have developed **e.g.** Dewria, Bardiha, Adhaura, Lohra, Dumrawan, **etc.** which are probably due to subsidence of the upper surface during Tertiary period.

8.3 SOUTHERN MODERATELY DISSECTED PLATEAU

This geomorphic region covers an area of 449 square kilometres (19.18 % of the total area) in the southern part of the Plateau. Its southern boundary is formed by the southern precipitous scarps up to the valley of Ausane river in the east. Ausane is the only major south-east flowing river which has cut deep inside the Plateau. The valley-walls which are in the form of steep scarps constitute the irregular boundary of this geomorphic region in the east. The northern boundary is also in the form of low scarps representing the valley-sides of the Goptha and Durgawati rivers. Part of the northern and north-western boundary is formed by 500 metres contour line. The western boundary runs along the State boundary of U.P. The region extends from 24° 33' to 24° 46' N. latitude and 83° 32' to 83° 56' E. longitude.

The highest elevations of this geomorphic regions are formed on south-western part where the Plateau is 641 metres high. The surface moderately slopes northwards and the height decreases to 511 metres on the opposite side along the boundary in the north. Along the longitude 83° 40' E. the height in the southern scarps zone is 609 metres and decreases to 571 metres along the same latitude near the northern boundary. Along 8°3 50' E. the height in the southern portion is 554 metres, which decreases to 472 metres near the tributary of the Durgawati river. The height along the eastern boundary near Goptha river valley is 458 metres and in the area enclosed by Gularia **khoh nala** it is 552 metres. This portion of the Plateau is a partially detached segment of the geomorphic region.

On the basis of recorded height it is clear that the whole geomorphic region slopes gently northwards. The slope increases towards the tributaries of Durgawati and Goptha rivers in the north-east.

The bedrock on the surface is hard sandstone over which the rivers have not been abled to carve deep valleys. All the rivers originating in the southern area flow in narrow channels northwards and form the tributaries of the Karamnasha flowing to the west, Durgawati flowing to the north and Goptha river flowing to the north-west. Some of the tributaries in the south-eastern parts join the Ausane river flowing to the south-east. since none of the rivers flowing within this region are master streams, deep valleys are rarely found. This enables us to separate this part of the Plateau as the southern moderately dissected Plateau.

In the south-western part of the Plateau we find herring-bone pattern in the tributary of Durgawati river while in the other parts the pattern is clearly dendritic.

The type of vegetation varies from area to area. In the south-western higher area open mixed vegetation is found which is not dense. Along the scarp margins dense mixed forest (mainly bamboo) is found. The type of vegetation changes in the lower eastern sections of the geomorphic region where dense mixed forest covers the area. The drainage network is somewhat fine. There are patches of open scrub vegetation interspersed in the dense mixed forest zone. These patches of open scrub vegetation are found mainly in depressed areas where settlements have developed.

The relative relief in the southern part near the scarp is below 20 m. but in the lower reaches of Karamnasha and in the middle reaches of Durgawati it is 60 to 80 metres which at some places increases to 80-100 metres. But in the deep notches of Durgawati it is more than 350 metres.

The dissection index in the 90 per cent of the region, in the western and central part is 0.1 to 0.2. In the Rohtasgarh area, west of Ausane river valley the index is 0.4 to 0.5 and in the Durgawati river valley area it is 0.5 . Thus only 10 per cent of this geomorphic region shows relatively high degree of dissection.

The area slopes northwards from the crest of the scarps and in the upper portion the volume of water is not sufficient to encourage rapid fluvial action. The drainage lines are, therefore, narrow and their erosive power is limited. Rainwater flows gently northwards. It is incapable to cut deep-valleys due to limited volume and limited velocity of water.

8.4 NORTH-EASTERN HIGHLY DISSECTED PLATEAU

The north-eastern segment of the Plateau forms a distinct geomorphic unit covering an area of 228 square kilometres representing 9.40 per cent of the total surface area of the Plateau. The region extends from 83^0 44' 30" E. longitude to 84^0 03' 30" and 24^0 40' N. latitude to 24^0 54' 30" N. This region may be called as north-eastern highly dissected Plateau. Its boundaries in all sides have been formed by steep scarps. The scarps to the south, south-east and northern boundary are steep and form a dividing line between the plain areas in the south and north. The boundary on the western part has been formed by the subsided valleys occupied by the Goptha river. The western portion of this geomorphic region is narrower while it is wider in the south and north-east. Development of Dumarkhar river valley has pushed the scarp boundary southward.

The highest elevation in this region is found in south central part where height is 534 metres. In the north-east the height gradually decreases to 413 and 322 metres which are the lowest heights in the north-eastern corner. This zone is drained by the Dhoba river which flows towards the north. The scarp zones in the east and in the north-west are higher in elevation and small **nalas** originate in these higher areas and join the Dhoba drainage system. Similar is the case in the other western half of the region. Due to the height of scarps many tributaries of Goptha river come flowing from the northern scarp area to join the Goptha in the south which forms the boundary of the region. The height along the northern scarp is 438 metrs and in the south along the Goptha valley it decreases to 321 metres. There are some of the areas scattered here and there which have heights ranging from 447 to 462 metres. The Banjari hill is the highest which is 530 metres high.

The drainage density is highest in the northern and north-western sections of this geomorphc region. There are very few rivers in a wide zone in the eastern part. The rivers coming from north to south are of varying length and size to join Goptha river. The main tributaries are Kurilwa, Piparkhar and, Chormara **nala.** The pattern in this part is dendritic. There is a dense network of tributary joining Dhoba river which flows to the north. The Dhoba river flows along the margin of the western scarps. Only one small river flows eastward which is known as Kachhuar **nala** in the upper part and Tutla river in the lower reaches.

In the south-central part of the Plateau in the Banjari hill area a radial pattern of drainage is clearly visible from this highest point the tributaries of Goptha flow from north-east to south-west, the Dhoba flows to the north and the tributarie of the Ausane river flow to the south-west.

Vegetation in the north-eastern corner is of open mixed type and in the rest of the area it is of dense mixed type. There are two depressed areas of small areal extent where the vegetation is open scrub type. In the north-western corner of the region also open mixed forest occurs.

Due to the development of a fine network of drainage lines the area has undergone high degree of dissectin. Relative relief is between 60 to 80 metres for the area as a whole but it increases from 60 to 100 metres towards the Goptha basin and in the Dhoba river basin the relative relief is 220 metres due to notches and **khohs.**

The dissection index in the Ausane divide and upper parts of Dhoba river is not much high but in the lower Dhoba basin the dissection index is 0.4 to 0.5. The Kalanwa river basin is a small basin but shows high degree of dissection.

The area has evolved on hard sandstone structure where fluvial processes have been active in carving out river valleys. Due to the high elevation of the scarp zone the rainwater flows towards the interior of the Plateau and unites in the west flowing Goptha, in the north flowing Dhoba. The valleys have a roughly north-south dimension in the north-western part and northwest-southeast alignment in the north-eastern part left of Dhoba river. There has been no major tectonic activitgy on this geomorphic region except along the subsided valleys of Goptha and Dumarkhar. The subsidence of the Dumarkhar basin has affected the alignment of the northern boundary of the scarps.

8.5 SCARP ZONE

As has been discussed earlier (Ch. 5.4) the scarp zone which forms the boundary of the Plateau towards the south, east and north is the most conspicuous and distinct feature of the Rohtas Plateau. The scarp zone covers an area of 328 square kilometres representing 14.04 per cent of the total Plateau area. This is why it is a separate geomorphic region. The scarps in the south begins from 83° 30' E. longitude and continue eastward in a gently sinuous alignment up to the valley of the Ausane river. In the valley area due to subsidence and headward erosion of the river the scarp zone takes a sharp turn towards the interior of the Plateau and its alignment in this part has become highly irregular due to its inward indentations. From the left bank of the Ausane valley the scarp again adopts its gently sinuous nature and continues up to the north-eastern corner where from Tarachandi (24° 55' 30" N. & 84° 03' 30" E.) it assumes a westward alignment. The scarp has penetrated deep towards the Plateau in the northern portion also where subsidence of the Dumarkhar basin has produced a southward indentation. From the valley of the Dumarkhar the scarp maintains a straight course up to Shergarh hill from where the area of subsided caves begin. All the north flowing rivers notably Durgawati, Goptha, Karsotha, Sura I, Sura II and Kohira have occupied subsided valleys and have produced deep notches and **khohs** along their entire courses. The Plateau has faced high degree of dissection due to the combined processes of subsidence and erosion. On account of this the alignment of the scarps which might have been straight or gently sinuous in the past has been completely destroyed. The scarps along the highly dissected and subsided sections are visible as high valley walls twisting and turning along the drainage basins of different rivers. Thus the scarps which might have been straight forming the northern boundary in the past have migrated finger like deep into the Plateau. In the west of Sura I, however, there has been no appriciable dissection of the main scarp face and they are as bold and straight as the scarps of the south.

The elevation of the scarp is highest towards the south-western section from where it gradually loses height eastward. Around the Ausane river basin the height is about 500 metres and it decreases further eastwards to 400 metres near the small river basin of Tutla. In the north-east Tarachandi the height of scarps slightly increases and is within 300 to 400 metres. The lowest height of the Rohtas scarps are in the zone of subsided valleys where they continue in a highly irregular course along the river valleys. The height in this section is below 350 metres. The north-western scarp is about 300 metres high.

The scarps have developed cutting across horizontal beds of limestone, shale and sandstone of which the sandstone provides the upper capping. Difference in character of rocks has affected the slope of the scarps in different parts. In the upper portion they are vertical at 90 angle but assume a cobra-hood type of alignment in the lower sandstone shale region. The geological variation of rocks has been responsible for the development of all types of slope e.g. convex, concave, and rectilinear.

The drainage throughout the scarps forms a parallel pattern. Although the main drainage lines of Plateau flow towards the north due to the general slope of Plateau from south to north there are a few streams who have been able to establish a north-south flow in a small area above the Plateau near the scarps. These drainage lines are not well developed. In the south they have not able to carve deep valleys on the steep scarp face. During rains the entire southern scarp face produces a waterfall like feature in the south.

In the north-eastern escarpment zone short streams parallel to each other are visible in large number. They are not well integrated drainage lines and simply drain out the water higher escarpment zone on the scarp face.

In the northern depressed areas the drainage specially along Sura I river has formed a herring-bone pattern, although, for the scarp zone area proper they look like parallel pattern.

The steep sides of the scarps are without any vegetation cover. They look like bare rocky surfaces. But in the lower sections of the scarps where the slope decreases different types of vegetation covers have developed. In the south-western portion open mixed bamboo forests are found. In the area left and right of the Ausane river a large patch of open mixed forest occurs. In the eastern and north-eastern parts open mixed vegetation covers the scarp zone which continues up to the valleys of Goptha river. A small patch of dense mixed *sal* covers the scarps left of Sura I river basin. In the remaining parts of the northern scarps open mixed vegetation occurs.

Relative relief in some parts of the scarps is mostly above 100 metres. It is highest in the south-western part where it is 300 to 360 metres. In the upper reaches of the Ausane it is 230 to 385 metres. In the eastern section it is 206 to 265 metres and in the northern half (eastern) it is 260 to 290 metres. The north-western part has a relative relief of more than 200 metres.

The dissection index in the southern and south-eastern scarp face is 0.4 to 0.5 because of varying height of scarps in different sections. Dissection index is highest in the Ausane river valley. In most of the scarp zone on the eastern and northern parts the dissection index is between 0.4 and 0.5 with larger areas showing more than 0.5.

Scarps were originally produced by block uplift of the Rohtas Plateau during the Tertiary period. Since then they have been modified by fluvial action. The irregular shape of the scarp in the northern part was produced during the block uplift of the Plateau which led to the formation of the subsided valleys.

8.6 SUBSIDED VALLEY

The subsided valley zones constitute one of the most interesting and conspicuous geomorphic region of the plateau. The total area covered by these valleys comes to 166 squre kilometres representing 7.09 per cent of the total Plateau surface. Subsided valleys are found in the south-eastern and northern areas of the Plateau. In the south-east there is only one region where the Ausane and its tributaries drain the subsided part. In the north there is a small depressed area lying equi-distant from Tarachandi hills in the north-east and shergarh hill in the west. The subsidence which occurred along 83°30' E. has produced an amphitheatre like feature which is drained by Dumarkhar river and its tributaries. The largest zone of subsided valleys is found in the north central parts of the Plateau where the Plateau surface is drained by Goptha, Durgawati, Karsotha, Sura I, Sura II and Kohira river. The valleys of all the rivers are inter-connected and from the finger-like extensions in the central body of the Plateau. The subsided areas has become wider in the north where the Durgawati and Karsotha systems meet.

The surface elevation in the subsided section at the head of Goptha river is 250 metres, at the head of Durgawati 320 metres, at the head of Karsotha it is 238 metres and at the mouth of Durgawati basin it is 103 metres. Thus we see that subsided valleys maintain a higher elevation in the interior of the Plateau at the head of main rivers. The elevation gradually decreases northwards which permits the river to flow from south to north. This coincides with the general slope of the Plateau from south to north. There are, however, abrupt changes of height at the head of the most of the rivers where the drainage lines leave the higher Plateau and enter into the zone of subsided valleys. As a result most of rivers have produced waterfalls whiles entering into the section of subsided valleys. The intensity of fluvial erosion in lower parts has dissected the area deeper and caused them to acquire lower elevation in comparison to the upper reaches of the subsided valleys where fluvial action is not so strong.

The drainage pattern in the Goptha, Durgawati, Koel, Karsotha, Sura II is dendritic. The Sura I river flows straight north and receives tributaries at right angles producing a herring-bone pattern. All the rivers flow towards the north. Durgawati river has a sinuous course of which Goptha is an important tributary flowing towards north-west. Karsotha has north-east direction and joins Durgawati at the edge of the Plateau. Since all the north flowing rivers are the tributaries of Durgawati they have produced a well developed dendritic pattern. In the south-eastern subsided valley occupied by the Ausane river also the pattern is clearly dendritic. The drainage network has developed in a basin which is almost circular in the shape and the pattern looks like a banyan tree.

The natural vegetation in the upper portion of the Ausane valley consists of dense mixed forest (mainly bamboo) and in the lower parts the vegetation is of open mixed like. In the northern valleys the lower elevation areas contain open mixed forest while in the upper reaches of the main rivers the vegetation consists of dense mixed forest with **sal as** the main spices.

The relative relief in the Ausane subsided valley in the south-east ranges from 260 to 385 metres. In the Goptha and Durgawati river valleys it is 260 to 350 metres but it decreases in the Karsotha valley where it is 230 metres. Sura I and Sura II show still lower little relief ranging from 100 to 240 metres.

The subsided valley area is a highly dissected area due to the presence of notches, **khohs** and straight valley walls. Dissection index in all north flowing river valleys as well as the south-east flowing Ausane is more than 0.5.

The evolution of this geomorphic region is complex. The initial features were produced by the subsidence of the upper sandstone strata in the underground caves which were produced by solution and draining out of limestone with the movement of underground water. The caves originated perhaps before the Tertiary period and the tectonic activity of this period led to the collapse of roof materials.

Fluvial action since Tertiary has modified the original subsided valleys and processes of headward erosion have produced notches along the valley walls. **Khohs** which can be seen today are continuing the cave formation which was active in the past.

8.7 DEBRIS SLOPE

The debris slope flanks the southern scarps between Kosdihra village in the south-west to the Ausane valley in the north-east. This is the only area where a debris slope has developed at the foot of scarps. The remaining parts of the scarps zone have a very narrow and negligible girdle of debris zone. The debris slope in the south covers an area of 124 square kilometres representing only 5.73 per cent of the study area. The highest elevation of this debris zone occurs along the foot of the scarps where the height is between 160 and 200 metres. The area slopes southwards and the height along the southern fringes is between 140 and 150 metres. The monotony of the slope is broken here and there by a few small isolated hills which rise at places from the general level. They are Tiara, Katudanr, Tenwa with height ranging from 165 to 188 metres.

The streams on the debris zone flow from north to south following the slope of the land. The tributaries are very short and make dendritic pattern in the higher areas of the debris slope but the main rivers joining the Son river have formed a parallel pattern. The streams are young and valleys are shallow and narrow.

The vegetation cover varies from west to east. In the south-west narrow patches of dense mixed forestt with bamboo are found. In the wider zone in the south the cover is provided by open mixed forest. In the small patches in the east open scrub vegetation occurs.

The relative relief in the higher elevation areas is between 20 and 40 metres but decreases towards the fringes. The dissection index in the upper portion is between 0.2 and 0.3 but decreases to 0.1 and 0.2 in the lower parts.

It is interesting to note that the debris slope is not distributed all along the foot of the escarpments. It is confined to the southern foot whereas the Plateau is bounded by scarps on the east and north as well. This may be probably due to the tilting of the Plateau from south to north that has determined the drainage pattern and texture. We have noted (Ch. 5.4.2) that the southern scarp face has mostly parallel small streams which do not develop a system as such. They come down from the Plateau cap, only during the rain, with weathered materials. Since the streams are not voluminous, they just carry the materials owing to high gradient, but as soon as they once on the foot, they become incapable to transport the materials further long. Even many streams do not exist thereafter. Therefore, we find a considerable wide belt of debris slope on the southern foot. The north and east flowing rivers, on the other hand, are comparatively mighty and long, and are capable to transport the weathered and eroded materials away from the foot of the escarpments. Therefore, the debris slope is insignificant on these areas.

8.8 ALLUVIAL TRACTS

An alluvial tract in between the scarp face and the Son is a separate geomorphic region having its own regional character. It covers an area of 192 square kilometres, about 8.62 per cent of the total study area. The alluvial fringe south of the debris slope is very narrow having a width of only half kilometre in the south-western and southern parts. The elevation in the narrow tract is 140 metres. From longitude 83^0 50' E. the alluvial tract increases in its width and has penetrated somewhat deeper in the subsided Ausane valley and Tutla river valley. The tract is widest near the Tutla river valley where it is 3 to 8 kilometres wide. The entire tract has an elevation below 140 metres which is the maximum height of the tract along the foot of the scarp in the north-east and along the foot of debris slope in the south. This slopes gently away from the scarps and the debris slope and along the fringe of the tract near Tutla valley on the left bank of the river Son the height is 118 metres. Thus the total variation of the height along the foot of scarp and the eastward fringe is only 22 metres.

Small streams occupy the alluvial tract flowing from north to south in the son. Almost all the streams originate at the foot of the scarps which forms a prominent water divide between north flowing major river and south flowing minor rivers on the alluvial tracts. The streams in their upper half have acquired dendritic shape while they form a roughly parallel pattern in the lower alluvial tract before joining the Son. North east of the Ausain Valley there has been no development of drainage line in the alluvial tracts. The only river which cuts across the tract is Tutla river which has not been able to reach up to the Son. The river dries up in the alluvial tract itself. The absence of integrated drainage lines in the north-eastern alluvial tract is probably due to the fact that the higher scarps do not permit water to flow towards the north-east or due to limestone structure. All the drainage lines originating from the north-east scarps flow towards the west away from the alluvial tract. There has been a development of a large number of rills in the north-eastern tract some of which are able to reach the son. But there has not been development of clear cut river valleys. The irregularity of drainage development is probably due to the extremely porous nature of the alluvial tract in this part which leads to the rapid seepage of rain water in the underground channels.

The original vegetation cover which might have been scrub or grass has been cleared from most of the tract. The area has been developed for agricultural purposes and now it forms an important agricultural zone in the study area. There are patches of open scrub vegetation on the isolated hills which rise from the level of the general alluvial plain. One notable area is Amarkha pahari where the remnants of former vegetation cover are present.

The relative relief in this area is below 20 metres because of the slight variation of height in the north-eastern, south-eastern boundaries of the alluvial tract. Dissection index is 0.1 because there has been no appreciable dissection of this geomorphic region.

As discussed earlier (Ch. 5.1) this alluvial tract has been formed by the combined processes of weathering and mass wasting of scarps and alluviation processes of the Son. The talus produced by the weathering and denudation of the scarp face was deposited at the foot of the scarps some of which were carried towards the fringes of the region. Over this deposit of talus and fragmented material derived from the scarp face, the son river has deposited finer clay materials which have burried the coarse materials below. This is a continuous process. Fragmented talus materials from the scarps have been continuously available to the tract over which the Son spreads a cover of finer materials specially during flood.

The soil profiles obtained from this region clearly show alternating beds of thick gravel and finer materials up to a depthe of 60 metres. Therefore, it can be inferred that the evolution of the tract has been initiated after the Tertiary uplift. The scarp might have receded backwards producing weathered materials which are spreading over the whole alluvial tract. The area has become suitable for cultivation only due to the deposition of alluvial material by the Son otherwise it must have remained in the same condition as the southern debris slope is found today. Therefore, it can be said that the Son has played a dominant role in the making of this geomorphic zone of economic importance.

9

CONCLUSION

Comprising an area of about 2340 square kilometres between latitudes 24° 30' 35" and 25° 00' 00" N., and longitudes 83° 20' 45" and 84° 05' 10" E. in the district of Rohtas of Bihar in India, the present study has attempted to analyse in detail all the terrestrial features of the Plateau's terrain on spatial and temporal scales and in terms of the processes which produced, sustain, and transform them within these scales. The discussion in the preceding chapters reveals that the Rohtas Plateau is a small museum of topographic features ranging from the tectonogenetic surfaces and forms of fluvial processes to residual and karst landforms. Most of the features are fascinating owing to their peculiar forms and materials that render their existence. It is astonishing that many such captivating features were left unrecorded, particularly of karst landforms. The foregoing chapters have successfully attempted to make geomorphological analysis of features of the Plateau as well as to make a record of all the unexplored features to fill up the gap in Indian geomorphology. A detailed discussion of all these aspects, e.g., escarpments, caverns, notches, waterfalls, plunge-pools, isolated hills, rocky-bumps and many other karst features has formed the broad theme of the present study.

Broadly speaking, geomorphology is the study of the landforms, their materials which make them and the processes that shape them. The discussion concentrates, therefore, on these three main aspects - *materials, processes* and *forms* endeavouring a complete geomorphic study of the Rohtas Plateau. The analysis of the *forms* spreads in five chapters. For *materials* and *processes* one each chapter is devoted. Thus the work is divided altogether into nine chapters. Each chapter includes various aspects of discussion to achieve the goal how the landscape of the Rohtas Plateau has been shaped out.

The foregoing chapters lead to the following conclusions -

1) The Rohtas Plateau comprises rocks of the Vindhyan formations (88.64 % of the study area) mainly limestone, shale and sandstone formed during the post-Cuddapah to late pre-Cambrian periods. The sedimentaries have been divided into two subdivisions, the lower Vindhyan and upper Vindhyan. The lower Vindhyan consists of marine calc-argillaceous sediments whereas the upper Vindhyans characterise almost wholly of red coloured sandstones and subordinates shale laid down in lacustrine and fluviatile environment. The thickness of the Vindhyans in the Rohtas Plateau has recorded the following bearings on the landforms :

a) The Kaimur sandstone which is 150 metres thick and overlies the Semri limestone works as a strong capping against denudational processes. The variety of landforms over the Plateau top is, therefore, limited. The hard red sandstones and quartzites are very resistant to erosion. Therefore, whatever fascinating landforms are recorded in this book, they are mostly due to tectonic disturbances and structural characteristics.

b) The limestone beds below the sandstones are liable to karstification. Water percolates through joints and cracks, but thick stratum of sandstone does not appreciate a good amount of percolation. That is why the caverns and the associated features which can be peeped through the lower parts of escarpments have taken long time in their development.

c) It is the great thickness of the sandstone strata that has preserved the numerous caverns even after the great Alpine earthmovements. Had these been a thinner capping, the whole Plateau mass would have been characterized with numerous collapsed karst valleys, most of them would have been dry.

d) There is a vast area for the development of karst landforms.The great thickness of the limestone strata has ample scope for further development.

2) The igneous activity in the Rohtas Plateau has minor geomorphic significance because it has occurred after the formation of Basal Stage, the lowest one of the Semri series. The activities have been recorded by the geologists as Porcellanite stage which is found only on the river bed of the Son in the form of vertical blades.

3) Of the structural characteristics in the Rohtas Plateau area folding is conspicuous in the Vindhyan rocks, which is clearly observable on the scarp face particularly in the limestone quarries. On the other hand the upper Vindhyan (Kaimur Sandstone) overlies unconformably and tilted.

The folding in the Lower Vindhyan rocks has geomorphic significance in the alignment of the limestone caves in the plateau. There are numerous caverns which are open towards the Son river and the northern plains. A close observation of numerous caverns reveals that the caverns are not at the same horizontal plane but they are following the arcuate paths and synclinal and anticlinal portions within the Plateau masses. For example, the Chunhatta caverns follow the synclinal paths whereas the Gupteshwar caverns follow the anticlinal portions. Thus the Lower Vindhyan rocks have multifolding system.

4) The major structural element of the Rohtas Plateau is the Great Boundary fault. There are a number of their subsidiary faults, parallel as well as oblique to the major fault but genetically connected with it. The faulting has resulted in the origin of a number of interesting features of geomorphological interest. The most impressive tectonic feature of the area is the Kaimur scarps from the top of which one has to descend along a sheer face for up to 400 - 500 metres. The scarps forcefully impresses any experienced geologist that faulting must have played a significant role in its genesis. The cross-faulting has originated some deep steep-sided river valleys which intrude into the plateau mass towards their source-heads and join the Son in the south and south-east. In the north too, such valleys originated by cross-faulting can be observed.

5) Rohtas Plateau has numerous joints which are produced during the Tertiary earth-movement. The joints of the Plateau have clear-cut impact on the development of drainage system of the area and thereby the evolution of the Plateau. Strikes and the relative density of the master set of joints seem to control the direction of some streams, their patterns and density in addition to the major control of underlying cavern system that has caused the origin of steep sided constricted and elongated valleys due to subsidence or collapse. The major joints run south-west to north-east and south-east to north-west. In all cases the arrangements of joints fits with the consequent and subsequent streams. The major rivers of the Rohtas Plateau follow the primary or major set of joints particularly in the upper reaches. Tributary streams seem to follow the secondary and minor joint-systems.

6) The study of soil stratigraphy reveals that the Rohtas Plateau consists of soils of *in-situ* origin for the most parts on the Plateau top. The hard red Vindhyan sandstones are resistant to weathering. Except in the depressions and **nalas,** drifted materials are rarely found. This indicates that the area has been under humid tropical climate in the recent geological past.

As regards the presence of sandstone and sandy clay and calcareous materials in the lowlying river valleys, particularly in the peripheral areas below the escarpments, the author is of the opinion that it is a case of collapse of former caverns. The sandstone angular boulders of large size are present on the river beds and banks. These materials cannot be drifted in nature, but they are collapsed materials. It suggests that the Plateau had numerous caverns in the geological past also before the Tertiary earth-movements. They have been collapsed during the Himalyan orogeny developing the narrow deep constricted deep valleys.

7) Both endogenic and exogenic processes have been responsible in originating and shaping the landforms of the Rohtas Plateau. Although the fascinating landforms are the result of tectonic movements through the ages, they are largely due to tectonics of the Tertiary era. The first epeirogenic movement occurred before the Cambrian which caused great boundary faults along the northern and southern margins of the basins and raised the Bhander beds above the base-level. This uplifted portion was subject to denudation and was totally/partly removed. The Caledonian earth-movements of the Silurian-Devonian period could not wrinkle the strata owing to their hardness but they could have raised the portion to a considerable height along the old faulted weak zones and consequently the Rewa series with some portions of the Bhander might have come above the base level of the old geosynclinal basin.

The period between the Caledonian and the Hercynian earth-movements was sufficient to reduce the Bhander and Rewa formations and the surface might have been lowered again. The Hercynian earth-movements of the Carboniferous-Permian periods which could originate the long and vast faulted basins over the Indian Peninsula might have affected this week zone and the Kaimur along with portions of the Rewa above and the Semri below which have uplifted throught the faulted margins. The period between the Permian and the Ecocene can be regarded as a silent age in regard to any epeirogenic or orogenic movements in the study area. There might have been, however, denudational processes which could remove the Rewa completely and the Kaimur partly. This period also witnessed karstification in the Semri series which are overlain by the Kaimur sandstone. It can be assumed that numerous karst landforms including large caverns might have developed. The time was sufficient for it. It can, however, be noted that the study area was not reduced to a peneplain as could be seen with the adjoining Chotanagpur portions. It remained high above the surface. It might have been planated.

The Tertiary earth-movement which was responsible for great Himalayan orogeny has scars of the epeirogenic movements over the Plateau and the adjoining Chotanagpur Highlands. The Rohtas Plateau was further uplifted along the boundary faults, it gained an ultimate height, the Semri was further thrown up above the basin level and numerous caverns collapsed resulting in separation of plateau portions from the main mass through karst valleys and development of long narrow steep sided deep valleys locally known as **khohs.** The exogenic processes including mainly weathering and mass-wasting, erosion and trasportation by running water and karstification by underground water have also played significant role in shaping the landforms (Ch. 3).

8) The character of drainage divides sometimes helps in the understanding of stage of landscape evolution. If the fluvial processes are working over a surface of homogeneous structure, the drainage divides are indistinct in the youth, very distinct in late youth and early maturity and again indistinct in the late maturity and old stage. If any thing is inferred about the Rohtas Plateau on the basis of these lines, it will be erroneous. The surface of the Rohtas Plateau is no doubt, homogeneous in character, mainly of Vindhyan sandstone but the underlying beds of limestone have very much influence on the development of the landforms. Therefore, on the basis of low water divides, the area is not in early or late maturity. The Plateau is still in youth stage.

9) The study of drainage basins of the Rohtas Plateau points out that the basins have uncommon shape particularly in regard to the length-breadth ratio. They have larger width than the length. The Kohira, Goptha, Durgawati, Ausane river basins are the clear-cut examples (Fig. 4.1). The Sura I, Sura II, Dhoba, Koel and Karsotha river basins seem to be elongated in shape but they are not actually separate river basins; and if Sura I and Sura II are considered together and the Karsotha and Koel basins are tagged with the Durgawati basin the width of the basin becomes larger than the length. This is a common feature with all the river basins over the Rohtas Plateau. This is a deviation from the general rule that a river has narrow catchment at the surface, it becomes wider in the middle and again it becomes narrower over a surface of homogeneous structure. The upper surface of the Rohtas Plateau has no doubt homogeneity in structure, that is the Vindhyan sandstone throughout the Plateau, but underlying beds of limestone and shale have left great impact on the development of streams and the shape of the basins. The underlying limstone beds have facilitated the development of intricate patterns of caverns and after the subsidence of cave roofs the river valleys have been formed haphazardly and owing to these subsided valleys the shapes of the basins have become uncommon and haphazard.

The drainage basins also highlight one common feature in all the basins and that is related to the gradient of main channels where we observe that almost every channel has moderate gradient in the upper course, it sometimes increases in the middle, it suddenly drops down with the steep slopes of waterfall followed by a plunge-pool at the base of the waterfall and flattened lower courses. It suggests that the lower portions of all these river basins have been subject to subsidence along the limstone caverns.

It also reveals that the upper reaches of the streams are discontinued. This may suggest that there is continued process of development of karst features.

10) It has been concluded in Chapter 4.2 that the catchment areas of the basins have not been determined only by the simple denudational processes of fluvial actions but there has been a vital role of the underground drainage resulting into the origin of a net of caverns and ultimately their collapse.

11) The drainage patterns are apparently guided by factors like joints, lithology, relief and eperogenetic movements over each of terrain. The most common patterns of drainage in the Rohtas Plateau are dendritic, pinnate and herring-bone. Besides these, rectangular, parallel, radial, barbed or fish-hook and other patterns are also observed in different localities. These have developed in different parts of the area in relation to the structural and physiographic control. Many individual streams of the area outline their individual patterns in the form of straight, meandering, rectangular and braided nature of their channels.

12) The study of hydrographic network (source-heads and confluences) of the Rohtas Plateau points out that the Plateau has a small percentage of the area under very high dissection. These areas of high density are physiographically a highly dissected Plateau lying between 300 m. and 600m. The Plateau is full of small and ephmeral rills and streams which have provided initiations and confluences together on the same unit. The Plateau is characterized mainly by dendritic, pinnate and herring-bone patterns of drainage. Had the density zones not been coincident, the patterns would be different from dominating dendritic one. Since most of the streams are small, most of the streams must be intermittent or short-lived in character.

The discussion of the drainage texture illustrates that it does not seem related to the stages of landform development, i.e., youth, maturity and old for fine, medium and coarse texture respectively as generalized sometimes; but the drainage texture of the Plateau is probably the result of rock structure, relief and vegetation and there is the overriding control of lithology.

14) The analysis of stream order concludes that the structure of the upper stratum of the Plateau is homogeneous throughout from where almost all the first and second order streams emerge. Therefore, the **Rb** for the first and second order streams is between 3 to 5. But the **Rb** between 3rd and 4th order streams and 4th and 5th order streams is above 5 which shows that streams enter into a different type of terrains on the lower reaches from a homogeneous terrain of the upper reaches of their courses. The lower parts of the rivers are flowing in elongated subsided valleys produced by collapse of upper sandstone strata. Almost all the rivers form waterfalls with sharp breaks of gradient and the **Rb** below the falls in almost all cases is high reaching to the highest figure of 14. It also concludes that drainage basins which have normal **Rb** are of common shape, i.e., like pear or apple. The drainage basins which have abnormal range of **Rb** in spite of homogeneous structure of the surface are mostly elongated, long and narrow. It also reveals that **Rb** goes beyound 3 to 5 on a homogeneous structure when there is abrupt rise or

fall of the surface. In the case of the basins of Plateau it happens because of the deep-cut river valleys, the valley sides of which amount sometimes to more than 90 of slope (Ch. 4.6).

15) The analysis of streams and their sinuosity indexes reveals that the reaches of the streams Koel, Goptha, Durgawati and Karsotha are not meandering. They are just sinuous. Other streams like Ausane, Dhoba, Sura I, Sura II, Lorsi and Kohira appear to have meandering courses but they have not developed flood plains in spite of their tortuosity. It also suggests that tortuosity is not of late mature or old stage, but it is exclusively of youth stage. The analysis of profiles drawn longitudinally and transverse to streams reveals the same because they show greater steepness in gradient. The hydraulic sinuosity index is not high in most of the river reaches. It shows that the rivers are not flowing on flood plain (Ch. 4.7). The study of the VI (valley index) and CI (Channel Index) reveals that most of the streams of the Plateau have deep-cut meandering beds due to collapse of materials. It also suggests that a considerable number of caverns was developed during the geological past which collapsed during the Tertiary movement. These meanders are not incised or entrenched over the earlier flood plain, but their development is quite different from rest of the Plateau rivers of Chotanagpur.

16) The study area includes a variety of surface features at different elevations. Important among them are the flood plain lying below 140 metres, notches and low-lying river-valleys, debris slope in between scarp and flood plain, steep and high escarpments, depressions and isolated residual hills, etc.

17) Tha analysis of relief measures (viz., relative relief, absolute relief, average relief, dissection index & ruggedness index, etc.) leads to the conclusion that most of the area of the Plateau is in upland, valleys are elongated, subsided, narrow and deep, and relief is very great over most of the area. The high dissection zones of the Plateau are not the absolute result of denudational processes related to the surface drainage but the phenomena associated with the karstification and later collapsing of caverns as well as tectonics have been responsible for defining such zones of high dissection. Therefore, the Plateau is, most probably between youth and maturity stage, better to say, it is progressing towards the late youth or maturity.

18) The analysis of slope gradient concludes that the Rohtas Plateau is such a region where slope of various inclinations ranging from monotonous alluvial tracts to the vertical escarpments is observed. It can also be concluded that the method devised by Raisz and Henry is more suitable and acceptable than that of Wentworth for the areal analysis of slope on the basis of contour from maps. The analysis of slope element concludes that they are not a product of erosion alone. To a large extent the elements have been introduced by tectonic activity. Erosional processes have added some elements to the original slope.

19) The analysis of the fascinating landscape of the Rohtas Plateau reveals that it presents a high land bounded by scarps. The deep-cut valleys with almost vertical walls on both the sides are confined to the Rohtas plateau. Other fascinating landforms are notches & **khohs,** duhi or rocky bumps, saucer shaped depressions, chuans, waterfall, water-kunds, apple shaped valleys, subsided valleys, isolated hills, flood plains and caves and their associated features. The scarps are very apparent to any casual observers. But many features are inaccessible. They require

additional enthusiasm as well as strong will; because they are located in the interiors where one can go only on foot. The caverns and their associated features require academic persuits as well. It is noted that the features of early stages in a karst region like sink-holes, dolines, uvalas etc. are not developed on the surface owing to the 90 metres thick capping of the old hard Kaimur sandstone but the percolation of water has facilitated the development of caves and the associated features. Such features are stalactite, stalagmite, dripstone, vadose and phreatic streams and depositional features both allochthonous and autochthonous. These are not easily accessible and observable to casual passers-by, but they are among the best examples of karst landforms in India. The author has tried, in spite of many odds, to explore the features. He is, however, humble to accept that many areas are left still unexplored.

20) The Rohtas Plateau has been divided into 8 geomorphic regions depicting the details of each regions. It is a fact that Rohtas Plateau is itself a geomorphic region, but within the delineation of the plateau there are variations in materials, forms and processes. The Plateau top has capping hard old red Vindhyan sandstone whereas the limestone-shale beds below are peeping through the scarps. On the foot are the debris slopes which are fringed by new alluvial tracts. Both endongenetic and exogenetic processes have evolved a variety of forms ranging from a near-flat alluvial zone to the high vertical rocky escarpments. In between these are the highly and moderately high dissected Plateaus, subsided valleys, notches and **khohs,** resistant outcrops in the form of ridges and hills, karst depressions and many other large and small details of topographic feature. The morphometric variables like drainage texture, hydrographic netwrok, relative relief, dissection index, ruggedness index, average slope and many other aspects of basin morphometry clearly depict the spatial variations within the Plateau.

Thus, summarising all these we may say that the Rohtas Plateau has been subject to crustal movements, it was uplifted during the Himalyan orogeny, it has reserves of fascinating landscape and it is a museum of karst landforms. Many of the feautres, particularly karst landforms are still unexplored and they require attention from and adequate number of enthusiastic geomorphologists. So, the study area has still large reserves of landforms to produce many theses like the present one.

BIBLIOGRAPHY

This BIBLIOGRAPHY includes all the books, articles and reports which have been cited in the text. It also includes such other references as have been consulted and found helpful in the preparation of this research work.

BOOKS

Ahmad, E. (1965) : *Bihar : A Physical, Economic & Regional Geography,* R.U. Ranchi.

Ahmad, E. (1972) : *Coastal Geomorphology of India,* Orient Longman.

Ahmad, E. (1973) : *Soil Erosion in India,* Asia Publishing House, Bombay, pp. 1-101.

Ahmad, E. (1976) : *Some Aspects of India n Geography,* Central Book Depot, Allahabad, pp.439.

Ahmad, E. (1979) : *Social and Geographical Aspects of Human Settlements,* Classical Publications, New Delhi.

Ahmad, E. (1985) : *Geomorphology,* Kalyani Publishers, Ludhiana.

Bloom, A.L. (1979) : *Geomorphology,A Systematic Analysis of Late Canozoic Landforms,* private Hall of Inida Pvt. Ltd., New Delhi.

Budel, J. (1973a) :'The Climatic Geomorphic System', *Climatic Geomorphology* (Ed. by E.Derbyshire), Macmillan, London, pp.104-130.

Budel, J. (1973b) :' The Climatogenetic Geomorphology, *Climatic Geomorphology,* (Ed. by Derbyshire), Macmillan, London, pp. 202-227.

Chanda, S.K. & Bhattacharya, A. (1982) : Vindhyan Sedimentation and Palaeogeography : Post-Auden Development, *Geology of Vindhyachal,* (Ed. by Valdiya, Bhatia & Gaur), Hindustan Publishing Corp., Delhi, pp. 88-101.

Cheng, Leong Goh (1971) :*Certificate of Physical Geography and Human Geography*, Oxford Univ. Press, Hongkong, pp.17

Chorley, R.J. (1972) :*Spatial Analysis in Geomorphology* (Ed.), Harper & Row, Publishers, New York, 393 pp.

Chorley, R.J. (1978) :Bases for Theory in Geomorphology, *Geomorphology,* (Ed. by Embleton et all) , Oxford Univ. Press, London.

Cooke, R.U. & Doornkamp, J. C. (1978) :*Geomorphology in Environmental Management,* Clarendon Press, Oxford, P. (Preface).

Cotton, C.A. (1960) :*Geomorphology : An Introduction to the Study of Landforms,* (whitecombe & Tombs, Christ-Church), 505 pp.

Cotton, C.A. (1969) :*Climatic Accidents in Landscape Making,* Hafner Pub. Corp. Inc., New York, p.354.

Dana, J.D. (1895) : *Manual of Geology*, Fourth Ed., New York, American Book Company.

Davis, W.M. (1909) : The Geographical Cycle, *Geographical Essays*, Ginn & Co., Boston (Reprinted 1954 by Dover Publications, Inc., New York).

Derbyshire, E. (1976) : *Geomorphology and Climate*, Ed. by E. Derbyshire, John Wiley & Sons, London, p. 1.

Doornkamp, J.C. King, C. A. M. (1971) : *Numerical Analysis in Geomorphology an Introduction*, Edward & Arnold.

Drew, Davod (1985) : Karst Prcoesses and Landforms, *Aspects of Geography*, (Ed. by Keith clayton and J.H. Johnson), Macmillan Education, London, pp.1-61.

Dury, G.H. (1970) : *Essays in Geomorphology*, Heinemann, London, p.173.

Dury, G.H. (1976) : *The Face of the Earth*, Penguin Books, Fourth Edition, London, p.9.

Easterbrook, D.J. (1969) : *Principles of Geomorphology*, McGraw Hill Book Company, New York.

Engeln, O.D.Von (1943) : *Geomorphology*, The Macmillan Company, New York, p.62.

Eyre, S.R. (1982) : *Vegetation and Soils*, Edward Arnold, Ltd., London.

Fairbridge, R.W. (1963) : Hypsometric Analysis, *The Encyclopedia of Geomorphology* (ed.) Reinhold Book Corp., New York,p.545.

Fairbridge, R.W. (1968) : Geomorphological Map, *Encyclopedia of Geomorphology*, (Ed .), Reinhold Book Corpn. New York, pp.388-403.

Fairbridge, R.W. (1968) : Plateau, *Encyclopedia of Geomorphology* (Ed .), Reinhold Book Corpn. New York, pp. 856-858.

Faniran, A. & Jeje, L.K. (1983) : *Humid Tropical Geomorphology*, Longman, London. p.1.

Ford,Derek, C. (1968) : Stalactite and Stalagmite, *The Encyclopedia of Geomorphology*, Ed.by R.W. Fairbridge, Reinhold Book Corp., New York, P. 1048.

Frank, Press & Raymond Siever (1978) : *Earth*, W.H. Freeman & Company, Sanfrancisco, pp. 126-127.

Gravelius, H. (1914) : *Flusskundeo : Goscensche Verlagshandlung*, Berlin, p.176.

Haines, H.H. (1978) : *The Botany of Bihar and Orissa*, Pt. 1, M/S Bishen Singh Mahendra Pal Singh, Dehradun, p.6.

Harry, Robinson (1969) : *Morphology and Landscape*, University Tutorial Press Ltd., London,p.149.

Hauwell, J.D. & Newman, M.D. (1973) : *Techniques in Physical Geography*, Macmillan, London, p.153.

Heilprin, Angelo (1890) : *The Principles of Geology*, Iconographic Encyclopedia Philadelphia p.7.

Herak, M. & Stringfield, V.T. (1972) : *Karst, Important Karst Regions of the Northern Hemisphere* ElsevierPublishing Co., London.

Holmes, A. (1970) : *Principles of Physical Geology*, Thomas Nelson and Sons Ltd. p.37.

James, P.E. (1949) : *A Geography of Man*, Latin America, 3rd ed., New York, p.88.

Khatri, Durga Prasad (1940) : *Bhutnath*, Lahri Book Depot, Kashi, pp. 1-110.

King, C.A.M. (1966) : *Techniques in Geomorphology*, Edward Arnold (Publishers) Ltd.,

London. p.31

Krishnan, M.S. (1968) : *Geology of India and Burma,* Higginbothams (P) Ltd., Madras.

Kumar, S. (1982) :Vindhyan Stromatolites and their Stratigraphic Testimony, *Geology of Vindhyachal* (Ed. by Valdiya, Bhatia& Gaur), Hindustan Publishing Corp., Inida, Delhi, p. 103.

Lake, P. (1958) :*Physical Geography,* 4th ed. (Ed.J.A. Steers), Cambridge Univ.Press, Cambridge.

Leopold, L.B. Wolman, M.G. & Miller, J.P. (1964) :*Fluvial Processes in Geomorphology,* Eurasia Publishing House (Pvt.) Ltd. New Delhi, p. 113

Lobeck, A.K. (1939) :*Geomorphology* : An Introduction to the Study of Landscapes, McGraw- Hill Book Co., New York, p.3.

Miller, A.A. (1953) :*The Skin of the Earth,* Methuen, London.

Moore, G.W. (1963) :*A Dictionary of Geography,* Penguin Book Ltd. Middle sex, England, p.16

Moore, G.W. (1968) :Limestone caves, *The Encyclopedia of Geomorphology* (Ed., by R.W. Fairbridge Reinhold Book Corp., New York, p.652).

Monkhouse, F.J. (1970) :*Maps and Diagrams,* Methuen & Co. Ltd., London, p.324.

Morisawa, M. (1985) :*Geomorphoglogy Texts, Rivers, Form and Process,* (Ed. by K.M. Clayton), Longman, London.

Mukherjee, P.K. (1973) :*A Textbook of Geology,* The world Press Pvt. Ltd., Calcutta pp. 328-332.

Mukhopadhyay, (1980) :*Geomorphology of the Subarnrekha Basin,* The University of Burdwan.

Mukhopadhyay, S.C. .(1982) :*The Tista Basin, A Study in Fluvial Geomorphology,* K. P. Bagchi and Co., Calcutta.

N.B.K. (1979) : *Combined Index to the New Book of Knowledge,* Grolier Incorporated, New York.

Oldham, T. (1983) : Manual of Geology of India,2nd Ed., Manager of Printing Govt. of India, Calcutta-5.

Ollier, C.D. (1979) :*Weathering,* English Language Agency & Longman Group Ltd., London, p.1

Pandey, P. (1970) :*Impact of Industrilisation on Urban Grwoth* (A Case Study of Chotanag pur), Central Book Depot, Allahabad.

Pitty, A.F. (1982) :*The Nature of Geomorpholgy,* Methuen London, p.1.

Powell, J.W. (1896) :*The Physiography of United States,* New York, American Book Co. Natl. Goeg. Soc. Mon.1. (Physiographic Proceses; pp.1-32; Physiographic features, pp.33-64).

Prakash, R.& Dalela, I. K. (1982) :*Stratigraphy of the Vindhyan in Uttar Pradesh* : A Brief Review, *Geology of Vindhyachal* (ed. by Valdiya, Bhatia & Gaur), Hindustan Publishing Corp., Delhi, pp. 55-79.

Prasad, A. (1973) :*Chotanagpur : Geography of Rural Settlements,* R.U., Ranchi.

Prasad, N. (1985) : Morphological properties of the Hill Side Slopes in the Barakar Basin and their Evolutioin,*Geographical Mosaic,* Ed. by S.C. Mukhopadhyay,Modern Book Agency Pvt. Ltd. Calcutta, pp. 137-

145.

Putnam, D.F. (1962) : Geography is a Practical Subject, *Geograhphy in the 20th Century* (Ed. by G. Taylor), Methuen, London, p.403.

Ray, S. (1978) :*Geomorphology of India,* Firma KLM, Ltd., Calcutta, pp.1-9.

Reiche, P. (1950) :*A Survey of Weathering Process and Products,* New Maxico University Publication, Geology, p.3.

Roy Choudhary, P.C. (1966) :*Bihar District Gazetteers,* Shahabad Secretariat Press, Bi;ar, Patna, pp.867-70.

Roy, A.K. & Bhattacharya A. (1982) :Regional Geomorphology of Vindyachal, *Geology .of Vindhyachal* (Ed. by Valdiya, Bhatia & Gaur) Hindustan Publishing Corp. Delhi, pp.9-22.

Sanders John E. (1981) :*Principles of Physical Geology,* John Wiley and Sons, New York, p.406.

Sarkar, S.N. (1968) :*Pre-Cambrian Stratigraphy and Geochronology of Peninsular In dia,* Dhanbad Publishers, India, pp.1-33.

Satpathi, D.P.P. (1981) :*An outline of Indian Geomorphology,* A Study in Regional Geomorpholgy of Singhbhum, Classical Publishing Co., New Delhi.

Singh, Kashinath & Kailashnath (1971) : *India* (Edited by R.L. Singh), National Geographical Society of Singh India, pp.612-643.

Singh, R.P. & Kumar, A. (1970): *Monograph of Bihar,* Bharati Bhawan, Patna.

Smith, Jacki (1984) : *Dictionary of Geography,* Arnold Heinemann.

Sparks, B.W. (1972) : *Geomorphology,* Longman, London,pp.1-4.

Stoddart, D.R. (1969) : 'Climatic geomorphology : Review and Re-assessment', *Progress in Geography* Vol.1, (Ed. by C. Board, R.J. Chorley, P.Hagget and D.R.Stoddart), Edward Arnold (Pub) Ltd., London, pp.163.

Strahler, A.N. (1971) :*Physical Geography,* Wiley Eastern Pvt. Ltd., New Delhi, p.415.

Sweeting, M.M. (1972) :*Karst Landforms,* The Macmillan Press Ltd., London.

Sweeting, M.M. (1968) : Karst, *The Encyclopedia of Geomorphology* (Edited by R.W. Fairbridge), Reinhold Book Corp. New York, p.582.

Thompson, G.A. & Talwani, M. (1964) :*Crustal Structure from Pacific Basin to Central Nevada,* J. Geophys, Res. I. 62 (22), pp. 4813-4837.

Thornbury, W.D. (1954) :*Principles of Geomorphology,* John Wiley and Sons, Inc., New York

Trefethen, J.M. (1960) :*Geology for Engineers,* East-West Press Pvt. Ltd., New Delhi, p.82.

Tricart, J. & Cailleux (1972) :*'Introduction to Climatic Geomorphology,* Longman, London, p.295.

Twidale, C.R. (1976) :*Analysis of Landforms,* John Wiley and Sons, Sydney, p.3.

Valdiya, K.S. (1982) :Tectonic Perspective of the Vindhyachal Regions, *Geology of Vindhyachal* (ed. by Valdiya, Bhatia and Gaur) Hindustan Publishing Corp., Delhi, pp. 23-30.

Wadia, D.N. (1961) :*Geology of India,* Macmillan and Co. bLtd., London, pp.126-137.

Whitten, D.G.A. Brooks J. R. V. (1972) :*A Dictionary of Geology,* Penguin Books.

Wilson, Lee (1968) :Slopes, *The Encyclopedia of Geomorphology* (Ed. by R.W. Fairbridge), Reinhold Books Corp., New York, pp. 1002-1020.

Williams, P.W. (1969) :*The Geomorphic Effects of Ground Water, Water Earth and Man,*

Ed. by Richard J.Chorley, Methuen, London, pp.269-284.

Wooldrige, S.W. & Morgan R.S. (1965) : *An Outline of Geomorphology,* (The Physical Basis of Geography) Longmans, London.

Worcester, P.G. (1965) : *A Text Book of Geomorphology,* Affliated East-West Press Pvt. Ltd., New Delhi, pp. 514-15.

Young, A. (1972) : *Slopes,* Longman, New York.

JOURNALS, MAGAZINES, & OTHER WORKS

Aghassy, J. (1970) : Jointing, Drainage and Slopes in a West African, Epeirogenetic Savana Landscape, Ann. of Assoc. of Am. Geog., Vol. 60, pp.286-298.

Ahmad, E. (1958) : Geomorphic Outline of Chotanagur, *Geog. Outlook*, R.U. Ranchi., Vol.2, pp. 16-22.

Ahmad, E. (1962) : Geomorphic Regions of Peninsular India, *Geog. Outlook*, R.U. Ranchi, Vol.1. pp. 1-20.

Ahmad, F. (1962) : The Origin and Significance of the Vindhyan Scarps, *Proc. 2nd, Symp., Earthquake Engg.* Univ. of Roorkee, pp. 513-525.

Ahmad, F. (1971) : Geology of the Vindhyan System in Eastern Part of the Son valley in Mirzapur District, U.P. *Rec. G.S.I.,* Vol. 96, Pt.2, pp.1-41.

Auden, J.B. (1933) : Vindhyan Sedimentation in the Son Valley, Mirzapur District, *Mem. Geog. Surv. India,* Vol. 62, Pt.2, pp.141-250.

Auden, J.B. (1943) : The Age of the Vindhyans, *Proc. 29th Ind. Sc.Cong.,* Baroda, pp. 111-112.

Bose, M.N. (1955) : Plant life in the Vindhyans, *Nature,* Vol. 178, pp.927-923.

Chhibber, H.L. (1948) : The Classification and origin of Plateaux with Special Reference to and Adjacent countries, *National Geog. Soc. Ind.,* Banaras, Bull.No.8, pp.1-12

Chhibber, H.L. (1950) : Easterly drift of the Son between Rohtas and Dehri, Bihar, Bull. No. 14, *National Geog. Soc. India.*

Chhibber, H.L. (1954) :The Drainage patterns observed in India and the Adjacent Countries, *National Geog. Soc. of India,* Bull.No. 20, Pt. 1, pp.1-116.

Chakraborty, S.C.(1965) : Notes on the Degradation of the Vindhyan Topography, *Geog. Rev. India,* pp.107-114.

Chatterjee, S.C. (1946) :Physiographic Evolution of Chota Nagpur, Calcutta, *Geog. Rev.* Vol. VII, No. 3-4.

Chorley, R.J. (1957) : Climate and Morphometry, *Journal of Geology,* 65, pp. 628-638.

Dangerfield, Cap.(1823) : *Memoirs of Central India,* by Sir John Malcolm.

Das, Sanjay (1977) : Outline of Methods of Detailed Exploration of Bedded Limestone deposits, with a reference to Amjhore Pyrites, *Rec. G.S.I.,* Vol. Pt. 2, pp. 14-37.

Datta, A.K. (1953) : Life in Vindhyan Period, *Science Cult.,* Vol. 18., No.7., p.329.

Davis, W.M. (1892) : The Convex Profile of Badland Divides, *Science,* 20, p.245.

Davis, W.M. (1913) : Meandering Valley and Underfit Rivers, *Ann. Ass. Am. Geog.,* Vol.3.

Davis, W.M. (1930) : Origin of Limestone Caverns, *Bull. Geol. Soc. Am.*, 41 (3), pp. 475-628.

Dunn, J.A. (1941) : The Economic Geology and Mineral Resources of Bihar Province, *Mem. G.S.I.*, Vol. 78.

Dutt, G.K. (1971) : Rohtas Plateau : A Regional Study, *Geog. Rev. India,* Vol. 33, No. 4, pp. 229-252.

Fair, T.J. (1947) : Slope form and Development in the Interior of Natal, *Trans. Geol. Soc.* S. Africa, 50, pp.105-20.

Fenneman, N.M. (1916) : Geology of Cincinnatic and Vicinity, Ohio, *Geol. Survey Bull.*, 19(4), p.202.

Fox, C.S. (1928) : Progress Report for 1928 (Unpublished report, *G.S.I.*).

Gupta, S.N.P. (1979) : Morphological Character of Hillside Slope in the Damodar Basin in Bihar, *Indian Geog. Studies,* No. 13 Geog. Res. Centre, Patna, pp. 16-21.

Horton, R.E. (1945) : Erosional Development of Streams and their Drainage Basins : Hydro logical Approach to Quantitiative Morphology, *Bull. Geol. Soc. Am.,* Vol. 56, pp.275-370.

Howell, B.F. (1956) : Evidence from fossils of the Age of the Vindhyan System, *Jour. Pal. Soc. India,* Lucknow, Vol.1, No. 1, pp.108-112.

Joffe, J.S. (1949) : Pedology, Pedology Publications, New Brunswick, New Jersey (Quoted in ' Soils and their Geographical Significance ' by D.F. Putnam in *Geography in the 20th Century,* Ed. by Tailor, p.223.

Jog, S.R. (1985) : Stream and Bifurcation ratios, *Trans. Inst. Indian Geographers,* Vol. 7, No.1, pp. 85-93.

Jones, M.S. (1909) : General Report on Vindhyan System, *Rec. Geol. Surv. Ind.*, Vol. 38.

Kedar Narain (1970) : Vindhyan Sedimentation in the Karwi Area, Uttar Pradesh, Rec. Geol. Surv. India, Vol. 98, Pt. 2,pp.96-106.

King, L.C. (1957) : The Uniformitarian Nature of Hill Slope, *Transactions of the Edinburgh Geol. Soc.,* 17, pp.81-102.

Krishnan, M.S. (1953) : The Structural and Tectonic History of India, *Mem. G.S.I.*, Vol. 81.

Krishnan, M.S. (1959) : The Great Vindhyan Basin of Northern India, Journal Geol. Soc.India, & Swaminath, J. Vol. 1 , pp. 10-36.

Kumar, Anil (1981) : Quantitative Geomorphology of Small Drainage Basins of Hazaribagh of Plateau, *Geog. Rev. India,* Vol. 43, No.3, pp.196-203.

Kumar, Anil & Pandey, R.N. (1981) : Morphometric Analysis of some Drainage Basins of the Hazaribagh Plateau Region, *N.G.J.I.,* Vol. XXVII, Pt.1 & 2, pp.50-59.

Lahiri, D. (1964) : Petrology of the Vindhyan Rocks around Rohtasgarh, India, *Jour. Sc.Petrol.,* Vol. 34, pp. 270-280.

Leopold, L.B. Langbein, W.B. (1966) : River Meanders, *Sc. Am.,* 214 (6).

Leopold, L.B. &Wolman, (1960) : River Meanders, *Bull. Geol. Soc.* M.G. Am., 71.

Louis, H. (1973) : 'The Problems of Erosion Surfaces, Cycles of Erosion and Climatic Geomorphology, *Climatic Geomorphology* (Ed. by E. Derbyshire), Macmillan London, pp. 153-170).

Mallet, F.R. (1869) : On the Vindhyan Series as Exhibited in the North Western and Central Provinces of India, *Mem.G.S.I.,* Vol. 7, Pt.1, pp.1-129.

Martonne, E.D. (1973) : Climate : Factor of Relief, *Climatic Geomorphology,* (Ed. by E. Derbyshire) Macmillan, London, pp.61-75.

Mathur, S.M. (1976) : A revision of the stratigraphy of the Vindhyan Supergroup in the Son Valley, Mirzapur Dist., U.P., *125th Year Celebrations of G.S.I.,* Misc. Pub. No.50, Symp. Bhopal, pp.7-20.

Mathur, S.M. (1981) : Cyclic Sedimentation in the Vindhyan Supergroup, *Rec. G.S.I.,* Vol.112, Pt.8, pp.1-6.

Medlicott, J.G. (1859) : On the Geological Structure of the Central portion of the Narbudda dist., *Mem. G.S.I.,* Vol.2, Pt.2.

Medlicott, H.B. (1859) : Vindhyan Rocks and their Associates in Bundelkhand, *Mem. G.S.I.,* Vol.1, Pt.1, pp. 1-95.

Melton, M.A. (1966) : Geomorphic Properties of Mature Drainage Systems and their Representaion in an E4 space, *Journal of Geology,* pp.35-54.

Mishra, R.C. (1948) : Evidence of the Existence of Life during Vindhyan Period, *Proc. 34th Ind. Sc. Cong.,* Delhi, p.144.

Mishra, R.C. (1949) : On Organic Remains from the Vindhyans (Pre-Cambrian), *Current Science,* Vol. 18, pp.438-439.

Mishra, R.C. (1969) : Vindhyan System, Presidential Address *Proc. 56th Ind. Sc. Cong.,* Pt.2.

Mukherjee S. (1979) : Slope Analysis of Brahamputra Basin in Nagaland, *Geog. Rev. Ind.,* Vol. 41, No.3, p.284.

Mueller, J.E. (1963) : An Introduction to the Hydraulic and the Topographic Sinuosity Indexes, *Annals Assoc. Am. Geog.,* Vol.58.

Oldham, T. (1859) : Remarks on the Classification of Rocks of Central India, Resulting from the Investigations of the Geological survey, *Jour. Asiatic Soc. Bengal,* Vol. 25, pp.224-256.

Oldham, R.D. Vrendenberg, E. & Datta, P. N. (1901) : Geology of the Son Valley in the Rewah State and of parts of the Adjoining Districts of Jabalpur and Mirzapur, *Mem. G.S.I.,* Vol. 31, Pt.1, pp.1-178.

Pandey, J.N. (1976) : Geomorphic Significance of Water falls of Kaimur Plateau (Bihar), Paper accepted for reading at the 63rd session of *Indian Sc. Cong.,* held at Waltair in Jan. 1976.

Pascoe, Sir. E.H. (1959) : A manual of the Geology of India and Burma, Vol.II, *G.S.I.,* pp.495-533.

Passarze, S. (1973) : Morphology of Climatic Zones or Morphology of Landscape Belts, *Climatic Geomorphology,* (Ed. by E. Derbyshire), Macmillan, London, pp.91-95.

Pathak, M. & Pandey, J.N. (1975) : Gupta Caves : A Geomorphic Study, paper accepted for reading at the 62 session of *Ind. Sc. Cong.,* held at Delhi, in Jan. 1975.

Pecsi, M. & Szilard, J (1968) : Principal Problems of Exploration and Nomenclature of Planated Surfaces, *Proc. of Symp. on Erosion Surfaces,* National Committee for Geog., Cal., pp.17-28.

Penck, A. (1905) : Climatic Features in the Land Surface, *Am. Jour. of Sc. Series,* 4,

19, pp. 165-174.

Penck, W. (1924) : Die morphologische Analyse; ein kapitel der physikal-ischen Geologic (Engelhorus Stuttgart, 283 pp; trans by H.Gzeh and K.C. Boswell, 1953, *Morphological Analysis of Landforms*, (Macmillan, London), 429 pp.

Pichamuthu, C.S. (1971) : Precambrian Geochronology of Peninsular India, *Jour, Geol. Soc., Ind.,* 12(3), pp. 262-73.

Pofali, R.M. & Hirekerur, L.R (1983) : Significance of Geomorphological Analysis for Soil Mapping, *Geog. Rev. Ind.* Vol.45, No.2, pp.24-34.

Prasad, N. (1977) : Stream Order Analysis of the Barakar Basin, *Indian Geographical Studies*, pp.69-73.

Prasad, N. (1978-79): Jointing Controls on Drainage in the Barakar Basin, *Geographical Outlook*, Vol. XIV, pp.101-105.

Prasad, N. (1979) : Hydrographic Netwrok and Drainage Basin Analysis : A Case Study of the Barakar Basin, *Geographical Rev. Ind.,* Vol.41, No.4, pp.297- 303.

Prasad, N. (1980) : Some Aspects of Meandering Streams of the Barakar Basin and their Sinuosity Indexes, *Geographical Rev. Ind.,* Vol.42, No.3, pp.296-299

Prasad, N. (1980) : Evolution of Slopes in the Barakar Basin, *The Indian Geographical Journal, Madras,* Vol. 55, No.1, pp.22-28.

Prasad, N. & Prasad,K. (1989) : Evolutionary Process and Forms in the Rohtas Plateau, Bihar, *Geog.Rev. of India,* Vol.51, No.1, Maroh, pp.37-47.

Prasad, N. & Prasad, K (1989) :Geomorphic Analysis of Surface Features of the Vindhyan Plateau in Bihar, *Landscape Systems and Eco ogical Studies,* Vol.12, No.1, pp.45-52.

Prasad N, & Prasad K. (1991) : Some Peculiarities of the Stream Hierarchy in the Karst Area of the Rohtas Plateau, Bihar, *Landscape System,* Vol.14, No.1, pp.82-85.

Rai, R.K. (1970) : The Study of Hillside Slopes in the Sonar-Bearma Basin, *Geog. Outlook,* Vol. VII, R.U. Ranchi.

Rai, R.K. (1979) : Some Aspects of Geomorphology of the Bewas Basin (M.P.), *Geog. Outlook,* Vol. XV, R.U. Ranehi.

Raisz, E. & Henry, J. (1987) : An Average Slope Map of Southern New England, *Geog. Review,* Vol.27, New York, pp.467-72.

Rawat, J.S. (1980) : River Sinuosity in Sona Basin : Its Significance in Landscape Analysis*Geog. Rev. Ind.*, 42, No.2, pp.181-82.

Robinson, A.H. (1948) : A Method for Producing Shaded Relief from Areal Slope Data, *Surveying and Mapping*, Vol.8, Washington.

Rode, K.P. (1945) : A new find Fossils in the Vindhyan Rocks of Rohtas Hills in Bihar, *Current Sc.*, Vol.15.

Sahni, M.R. (1962) : The Vindhyan System of India, *Rec. G.S.I.,* Vol. 91, Pt.2, pp. 271-278.

Savigear, R.A.G. (1960) : Slopes and Hills in West Africa, *Zeit, Fur, Geomorphology,* Suppl. 1/Morphologie des Versant, pp. 156-71.

Scheidegger, A.E. (1965) : The Algebra of Stream - Order numbers, *U.S. Geol. Surv.*, Prof. Paper 525-B, pp.187-189.

Shreve, R.L. (1967) :Infinite Topographically Random Channel Network, *Jour. Geol.*, 75 (2), p. 178-186.

Singh, R.L. (1974) : Morphometric Analysis of Terrain, *National Geog. Soc. Ind.*, Varanasi, Bull. No. 22, pp.1-24.

Singh, R.P. (1958) : Structure, Drainage and Morphology of Chotanagpur Highlands, *Geog, Outlook* R.U. Ranchi, Vol. 2, pp.31-40.

Singh, R.P. (1969) : Geomorphological Evolution of Chotanagpur Highlands, India, *National Geographical Soc. Ind., Varanasi.*

Singh, S. Singh,. (1992) : Morphogenesis of Cavernous Topography in Sub-humid Tropical
M.S. Dubey, A Environment - A case study of Sauptadhdam Cave, India, *Z.Geomorph. N.F., Suppl. Bd.*85, pp.127-144.

Smith, G.H. (1935) : The Relative Relief of Ohio, *Geog. Rev.*, Vol.25, pp.275-284.

Smith & (1952) : Hypsometric (area altitude) Analysis of Erosional Topography, *Bull.*
Strahler, A.N. *Geol. Soc. Am.*, Vol.63,00. 1117-1142.

Smith & (1956) : Quantitative Slope Analysis, *Bull. Geol. Soc. Am.*, Vol. 67.
Strahler, A.N.

Swinnerton, A.C. (1932) : Origin of Limestone Caverns, *Bull. Geol. Soc. Am.*, 43 (3), pp.663-693.

Vredenberg, E. (1906) : Suggestions for the Classification of the Vindhyan System, *Rec. G.S.I.*, Vol.33.

Wentworth, C.K. (1930) : A Simplified Method of Determiningthe Average Slope of Land Surfaces, *Am. Jour. Sc.*, series 5, Vol. 20, (New Haven, Conn.)

Wood, A. (1942) : The development of Hillside Slope, *Proc. Geol. Ass.*, 53, pp. 128-40.

UNPUBLISHED WORKS AND THESES

Pandey, J.N. (1978) :*Geomorphic Study of Kaimur Plateau Bihar,* unpublished Ph.D. Thesis, Magadh Univ., Bodh Gaya.

Prasad, K. (1972) :*Geomorphology of the Kaimur Plateau in Rohtas District*, Master Thesis, Ranchi University, Ranchi.

Prasad, N. (1979) :*Geomorphology of the Barakar Basin,* Unpublished D.Litt. Thesis, Ranchi University, Ranchi.

Prasad, Shambhu (1983) :*Industrial Resources Development & Prospect in Rohtas Plateau Region*, Master Thesis, R.U. Ranchi.

MAPS CONSULTED

Krishnan, M.S. (1953) : Movements in India During the Tertiary, The Structural and Tectonic History of India, *Mem. G.S.I.*, Vol. 81, p.59.

Lal, B.B. (1985) : Location Map of Rohtas District, District Census Handbook, Bihar Govt. Press, Patna, p.2.

Roy, A.K. & (1982) : Geological set-up Map of the Vindhyachal, *Geology of Vindhyachal*

Bhattacharya, A : Ed. by Valdiya, Bhatia & Gaur, Hindustan Publishing Corp. (India) Delhi, p.11.

Surveyor General (1971-onward) of India, Survey of India Office, Dehradun (U.P.) :Topographical Sheets of the Rohtas Plateau and its Adjoining Areas
1:50000 Topo-Sheets
63 P/5, 6, 9, 10, 13, 14;
72 D/1, 2

9788185891057